Immobilized Microbial Cells

Immobilized Microbial Cells

K. Venkatsubramanian, EDITOR

*H. J. Heinz Company and
Rutgers University*

Based on a symposium

jointly sponsored by

the ACS Divisions of

Microbial and Biochemical Technology,

Agricultural and Food Chemistry,

and Carbohydrate Chemistry

at the 176th Meeting of the

American Chemical Society,

Miami Beach, Florida,

September 14, 1978.

ACS SYMPOSIUM SERIES **106**

AMERICAN CHEMICAL SOCIETY

WASHINGTON, D. C.   1979

Library of Congress CIP Data

Immobilized microbial cells.
(ACS symposium series; 106 ISSN 0097–6156)

Includes bibliographies and index.

1. Industrial microbiology—Congresses. 2. Microorganisms, Immobilized—Congresses. 3. Micro-orga-
nisms, Immobilized—Industrial applications—Congresses.
I. Venkatsubramanian, K., 1948-   . II. American
Chemical Society. Division of Microbial and Biochemical Technology. III. American Chemical Society. Division of Agricultural and Food Chemistry. IV. American
Chemical Society. Division of Carbohydrate Chemistry.
V. American Chemical Society. VI. Series: American
Chemical Society. ACS symposium series; 106.

QR53.I45   660'.62   79-15794
ISBN 0-8412-0508-6   ACSMC8   106 1-258 1979

ACS Symposium Series

M. Joan Comstock, *Series Editor*

FOREWORD

The ACS SYMPOSIUM SERIES was founded in 1974 to provide a medium for publishing symposia quickly in book form. The format of the Series parallels that of the continuing ADVANCES IN CHEMISTRY SERIES except that in order to save time the papers are not typeset but are reproduced as they are submitted by the authors in camera-ready form. Papers are reviewed under the supervision of the Editors with the assistance of the Series Advisory Board and are selected to maintain the integrity of the symposia; however, verbatim reproductions of previously published papers are not accepted. Both reviews and reports of research are acceptable since symposia may embrace both types of presentation.

CONTENTS

PREFACE

B iochemical processing with immobilized microbial cells represents a novel approach to biocatalysis. Such a system offers a number of unique advantages over traditional fermentation processes as well as the more recent immobilized enzyme processes. Although this concept is still relatively new, a few immobilized cell systems have already been commercialized. This, in turn, has triggered a surge of research activity in this exciting and rapidly growing field. Numerous conferences and symposia have been held on the subject of enzyme engineering in recent years. Although they contain a few papers on the subject of immobilized microbial cells, no single conference was devoted to covering this subject matter exclusively. Therefore, we organized a symposium on immobilized microbial cells as part of the 176th Annual Meeting of the American Chemical Society held at Miami Beach in September 1978.

This volume contains most of the papers presented at the symposium. In addition, several chapters written by leading experts in the field have also been included. Several important aspects of immobilized microbial cell technology are discussed here: carriers for immobilization, methods of cell attachment, biophysical and biochemical properties, reactor design, and process engineering of bound cell systems. A number of applications in the food, pharmaceutical, and medical areas—including those commercialized already—have been described. In essence, this is a comprehensive single volume state-of-the-art presentation of immobilized microbial cell systems.

The first chapter by Vieth and Venkatsubramanian provides a broad overview of the subject matter including the rationale for immobilizing microbial cells, the advantages and disadvantages of such an approach, and the overall prospects and problems of a technological development based on bound cell systems. The chapter by Messing and associates discusses the critical pore dimensions needed for fixing microorganisms inside various inorganic matrices. This is followed by an interesting discussion on the adhesive forces that come into play in fixed microbial systems.

A series of biochemical processes mediated by fixed cells are described next. They vary in complexity in terms of the number of individual enzymatic reactions, and coenzymes involved. Included in this section are descriptions of immobilized cell systems for producing coenzyme A, pantothenic acid, antibiotics, and extracellular enzymes. In addition, waste treatment applications such as phenol degradation and denitrification are outlined.

Several important industrial applications are discussed next, starting with two commercial processes for the conversion of dextrose to fructose. The chapters by Bungard and co-workers and Roels and his associates describe two different approaches to this interesting commercial problem. Because of the commercial importance of this process, we have also included a paper by Goldberg on the use of glucose isomerase enzyme (as opposed to the whole organism containing the enzyme immobilized on a porous polymeric matrix). Chibata discusses several industrial applications of immobilized microbial cells as practiced in Japan. The chapters by Mattiasson and Suzuki and his associates discuss many interesting analytical applications of immobilized cell systems. The final chapter by Kastl describes a process for immobilizing isolated organelles and use of such a system in detoxifying drugs.

I am indebted to all the authors for preparing the manuscripts to meet a tight publication schedule, and to the reviewers for their prompt responses. Many thanks are due to Charles Cooney and George Charalambous of the Microbial and Biochemicl Technology Division and the Agricultural and Food Chemistry Division, respectively, for encouraging me to organize this symposium, and to John Whittaker for serving as cochairman of the symposium. I am thankful to the ACS Books Department for its assistance. Finally, the impeccable secretarial help of Diane Otto is gratefully acknowledged.

H. J. Heinz Company   K. VENKATSUBRAMANIAN
Pittsburgh, Pennsylvania
April 10, 1979

Immobilized Microbial Cells in Complex Biocatalysis

WOLF R. VIETH and K. VENKATSUBRAMANIAN[1]

Department of Chemical and Biochemical Engineering,
Rutgers—The State University, New Brunswick, NJ 08903

Continuous heterogeneous catalysis by fixed microbial cells
represents a new approach to established fermentation processes.
Immobilization of isolated (and purified) enzymes and microbial
cells mediating simple, monoenzyme reactions has already been
reduced to industrial practice. However, the development of
immobilized cell systems to carry out complex fermentation pro-
cesses--characterized by multiple reactions and complete reaction
pathways involving coenzymes--is still in its infancy. Drawing
upon our rather concerted effort in this area over the past several years, we are appraising the prospects and problems of such
a technological advancement in this brief communication.

The Approach

In earlier papers from this laboratory, we have proposed the
terms "Controlled Catalytic Biomass" and "Structured Bed Fermentation" to describe immobilized cell systems effecting complex
biocatalysis (1,2). The meaning of these terms is obvious when
one considers the biocatalyst in relation to its microstructure,
predesigned catalytic reactor design, and controlled catalytic
activity vis-a-vis cellular reproduction. Some of the potential
advantages of such a catalytic system are summarized in Table I.
Examining the character of microbial cells in classical
fermentation, it is clear that they possess the desired catalytic
machinery in a highly structured form. The controlled conditions
of fermentation permit retention of this meticulous structural

---

[1]Also with: H.J. Heinz Company, World Headquarters, P.O. Box 57,
Pittsburgh, Pennsylvania 15230.

Presented at the Symposium on "Immobilized Cells and Organelles,"
ACS National Meeting, Miami Beach, September, 1978.

TABLE 1

POTENTIAL ADVANTAGES OF IMMOBILIZED WHOLE CELL SYSTEMS
OVER CONTROLLED FERMENTATIONS

1. Placement of Fermentation on Heterogeneous Catalysis Design
    Basis

2. Higher Product Yields

3. Ability to Conduct Continuous Operations As Opposed to
    Traditional Batch Fermentation

4. Operation at High Dilution Rates Without Washout

5. Ability to Recharge System by Inducing Growth and Reproduc-
    tion of Resting Cells

6. Decrease or Elimination of Lag and Growth Phases for
    Product Accumulation Associated With the Non-Growth Phase
    of the Fermentation

7. Possibility of Accelerated Reaction Rates Due to Increased
    Cell Density

integrity but the resulting cellular suspensions are usually at
low concentration. Considering free enzymes derived from these
cells, it is possible to concentrate them by extraction processes,
but lacking the ancillary structure which stabilizes them in the
cell, they are relatively unstable. Some structural reconstitu-
tion is possible by immobilization, leading to higher concentration and better stability but one is then restrained to consideration of single step or two-step reactions. With immobilized
cells, one has the concentrated form, there is structural pre-
servation and stability together with the possibility of improved
reactor design, based upon the characteristics of the carrier.
Thus, immobilized cell systems constitute an important option
within the framework of biochemical technologies (Table 2).
The overall rationale for whole cell immobilization is outlined
in Table 3.

In all our work, we have used reconstituted bovine hide
collagen as the carrier matrix of choice. The biomaterial,
collagen, offers a number of unique advantages as a support for
microbial cell immobilization. Other publications from our
laboratory describe these advantages as well as the procedures
to prepare fixed cells in detail (3, 4). We have attached many
different microorganisms in this manner; some of the complex
reactions mediated by such fixed cell preparations are shown in
Table 4.

Process Variables

Several important considerations in the preparation and use
of collagen-bound cell systems are adumbrated here with citric
acid production by immobilized Aspergillus niger as an example.
The collagen membrane must be crosslinked to make it structurally
strong enough to withstand the shear forces in reactor operation.
It was found that post-tanning the collagen-cell membrane by
exposing it to a 5% glutaraldehyde solution for one minute resulted in an optimal retention of catalytic activity
which was a linear function of the cell loading. We can load
the structure up to 70% cells (by dry weight) and the amount of
expressed activity in batch assay increases proportionately.
However, the mechanical strength drops off too drastically, and a
good compromise is 50% cells on a dry weight basis. In the
course of these studies, we came to realize that the dehydration
of cells is deleterious; even under refrigerated conditions cell
activity could reduce significantly. This has led us to new
dispersion techniques and/or drying or solidification techniques
to preserve these fragile structures which can so easily denature
(6).

Maximal catalytic activity of the cells is retained upon
immobilization when the cells are in the proper physiological
state. This corresponds to an optimal induction of enzyme
activities participating in the desired reaction sequence;

TABLE   2

BIOCONVERSION NETWORK

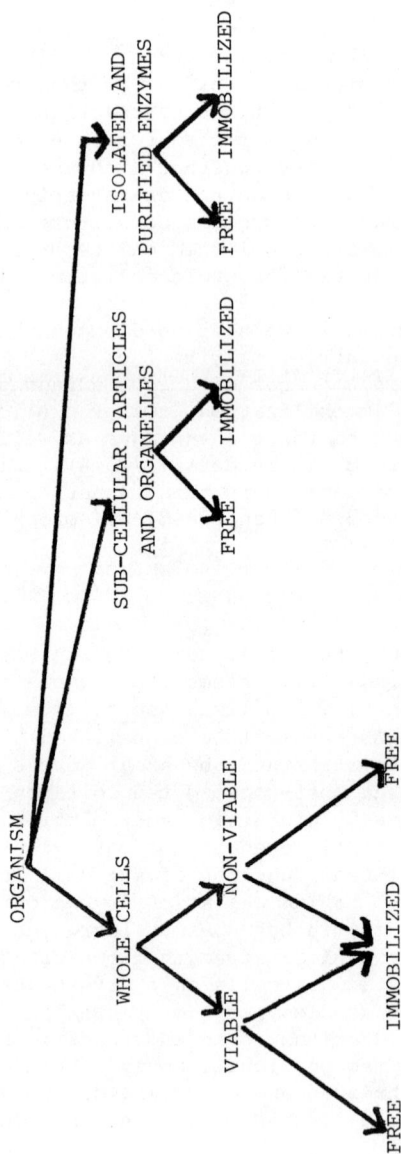

TABLE 3

RATIONALE FOR WHOLE CELL IMMOBILIZATION

1. Obviates Enzyme Extraction/Purification

2. Generally Higher Operational Stability

3. Lower Effective Enzyme Cost

4. High Yield of Enzyme Activity on Immobilization

5. Cofactor Regeneration

6. Retention of Structural and Conformational Integrity

7. Greater Potential for Multi-Step Processes

8. Greater Resistance to Environtal Perturbations

TABLE 4

COLLAGEN-IMMOBILIZED CELL SYSTEMS

	Microorganism	Substrate	Product	Comments
1.	Serratia marcescens	Glucose	2-Keto gluconic acid	Multi-enzyme
2.	Acetobacter sp.	Ethanol	Acetic acid	Multi-enzyme; cofactor
3.	Corynebacterium lilium	Glucose	Glutamic acid	Pathway (primary metabolite)
4.	Aspergillus niger	Sucrose	Citric acid	Primary metabolite
5.	Chloroplast	Water	Oxygen	Immobilized organelle; first step in biophotolysis of water
6.	Anacystis nidulans	Water	Oxygen	Immobilized algal cells
7.	Anacystis nidulans	Nitrate	Ammonia	Biological nitrogen fixation
8.	Streptomyces griseus	Glucose	Candicidin	Antibiotic synthesis; secondary metabolite
9.	Pseudomonas aeruginosa	---	---	Concentration of plutonium from waste waters (bioadsorption)
10.	Klebsciella pneumoniae	Nitrogen	Ammonia	Microbial fixation of atmospheric nitrogen
11.	Mammalian erythrocyte	---	---	Model studies of in vivo enzyme action

it is manifested in peak product synthesis rate in the fermentation. For citric acid production with A. niger, it turns out to be 72 to 96 hours in batch fermentations. Of course, in a typical fermentation process one has to repeat this pattern each time. A better alternative, it would seem, would be to harvest the cells at their peak activity, followed by their immobilization so as to retain them in a viable state for reuse until their stability has decreased to an uneconomical point.

Once immobilized, the cells must be kept in a viable state in the membrane without further excessive reproduction. This is necessary to channel the substrate into the desired product rather than to additional cell mass. Besides, it would minimize cell elution from the carrier matrix as well as preserve the mechanical integrity of the carrier. We have found that one way to accomplish this is by limiting the concentration of one of the essential nutrients in the medium; for example, nitrogen concentration. An indirect benefit of this approach is lowering the growth of contaminating organisms.

Ease of reactor scale-up is an important process engineering consideration; maximizing the efficiency of contact between the catalyst and its substrate is an equally critical issue. We have determined that where the bound-cell membrane can be rolled into a spiral wound reactor configuration (6), it provides excellent contact efficiency. The collagen membrane is wound together with a polyolefin Vexar spacer material. The resulting open multichannel system promotes plug flow contact with very low pressure drop even when operating with particulate substrate matter which would cause plugging problems in the conventional type of fixed bed operation. Fermentation substrates are often characterized by precisely this type of substrates; so this is a large plus factor in favor of this type of design. Furthermore, it is possible to design-in high activity per unit volume, as a result of the coiling of a large amount of membrane into a confined volume. The basis for scale-up becomes then simply the membrane surface area.

Presented in Figs. 1 and 2 are data relating to external and internal mass transfer for the case of citric acid synthesis. The effect of linear velocity on the observed reaction rate (Fig. 1) shows, for this case, the presence of a significant boundary layer resistance below a flow rate of 235 ml/min. The existence of non-negligible pore diffusional resistance is deducible from Fig. 2, in which the dependence of observed reaction rate on film thickness is depicted. Overall the immobilized cells exhibited about 50% of the specific activity of the free cells (in fermentation) toward the production of citric acid.

With regard to other significant factors, oxygen transfer can be singled out as of paramount importance. To enhance this transport step, we operated the spiral wound reactor countercurrently. In other words, a special provision was incorporated into the reactor design to allow flow of pure oxygen countercurrent

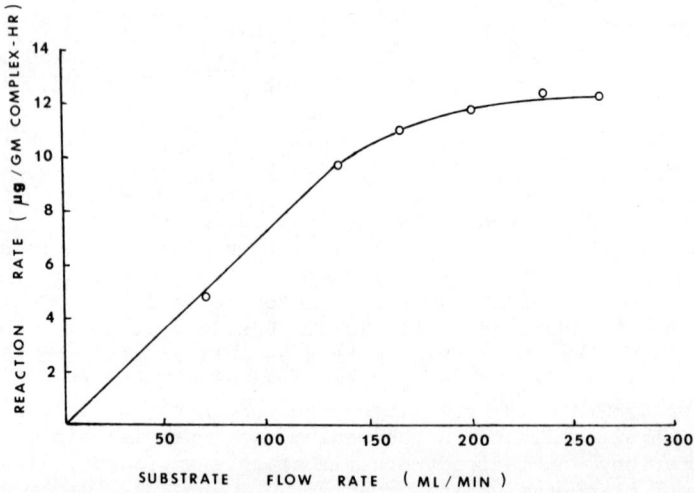

Figure 1.   Dependence of reaction rate on linear velocity

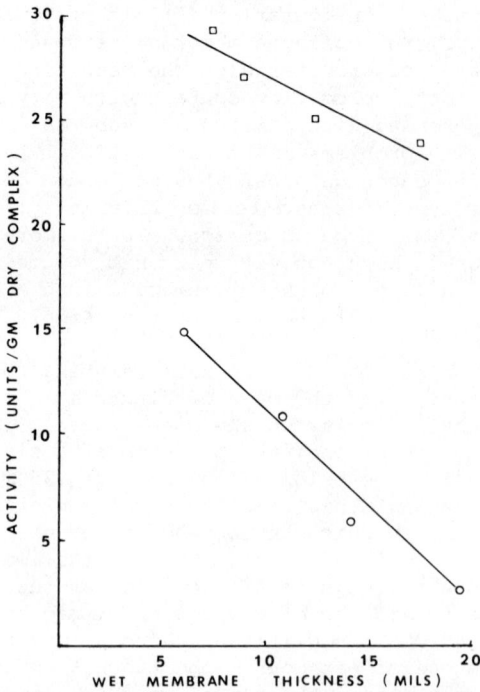

Figure 2.   Effect of membrane thickness on citric acid production rate. (○) Shake
flask, (□) reactor.

to the flow of substrate; in this sense, the overall system
operated as a combined absorber-reactor. Dissolved oxygen concentrations of 80 to 90% of saturation value were maintained
throughout the course of the reactor runs. Referring back to
Fig. 2, the increased specific activity of the catalyst observed
in the reactor compared to that in a shake flask is attributable,
at least in part, to improved oxygen transfer in the reactor.
Thus, the simple, effective, flexible use of the membrane form
in this type of reactor has demonstrated several additional
positive features.

From a practical standpoint, the two most important characteristics of an immobilized cell catalyst are its activity and
its operational stability. The latter parameter is usually expressed in catalyst half-life. The amount of activity, say in
International Units (I.U.), would be a function of cell-to-
carrier ratio. As mentioned earlier, a 50% loading ratio has been
found to be optimal.

Aspergillus niger cells attached to collagen exhibit good
activity retention, as shown in Table 5. Please note that rate
comparisons have been made on the basis of maximal rate. If one
uses an integrated average rate obtained over the entire period
of the batch fermentation cycle, the comparison becomes even
more favorable for the immobilized cell system, since it experiences a very small lag period preceding citric acid synthesis.
In addition to specific productivity rates, it is also necessary
to examine the relative concentrations of the product in both
cases, as the titer value is very crucial with regard to product
isolation and purification. Data obtained thus far indicate
that bound cells yield 8 to 40% of the final concentration
obtainable in fermentation. Half-life of the catalyst was 138
hours.

Chromatographic analysis of reaction products of citric acid
synthesized by fixed cells reveals the presence of products
generated from side reactions. They include isocitric acid,
oxalic acid and trace quantities of gluconic acid. Isocitrate
is perhaps the major one, amounting to as much as 15 to 20% of
citrate. Oxalic acid formation in citric acid fermentations is
reported to be dependent both on pH and on the extent of aeration.
By proper control of pH and dissolved oxygen levels, it might be
possible to reduce the formation of oxalate.

Conclusion

Immobilized cell and organelle systems offer a great deal of
promise in mediating many reaction schemes to produce commercially
important products.

Unlike bound mono-enzyme systems, catalysis by fixed cells
is quite complex and many basic aspects are yet to be understood.
However, technical feasibility of rather elaborate immobilized
cell processes, as exemplified by citric acid production through

TABLE 5

CITRIC ACID SYNTHESIS BY IMMOBILIZED CELLS

Sample	Maxiumum specific productivity (g acid/g dry cells-h)	Relative maximum specific productivity (%)
Fermentation	0.0043	100
Resting Cells	0.0045	104
Immobilized Cells	0.0021	48.4

Fermentation data obtained from 5-ℓ stirred fermentor; others from shake flasks. Sucrose at 40-ℓ was used as the substrate in all cases.

intact function of the TCA cycle enzymes, has been demonstrated. Investigation of the basic problems of structured-bed fermentation systems (cell physiology, cell viability, transport resistances, oxygen transfer, microbial contamination) is now being pursued in our current work. Perhaps the greatest potential for immobilized cell systems lies in replacing complex fermentations such as secondary metabolite production. Some of the further developments in this field should clearly be steered in this direction.

Acknowledgements

The authors are grateful to the contributions of Mr. Charles Bertalan; some of the results reported here have been drawn from his thesis. Financial assistance for our work in this area was provided in part by a National Science Foundation Grant (AER 7618816), Ethyl Corporation and H.J. Heinz Company.

Literature Cited

1. Venkatasubramanian, K., and Vieth, W.R., Progress In Industrial Microbio. (in press).

2. Vieth, W.R., Annals New York Acad. Sci. (in press)

3. Vieth, W.R., and Venkatasubramanian, K., Methods Enzymol. 34, 243 (1976).

4. Venkatasubramanian, K., Vieth, W.R., and Constantinides, A. in E.K. Pye and H.H. Weetall (Editors), Enzyme Engineering, Vol. 3, Plenum Press, New York (1978) pp. 29-42.

5. Vieth, W.R., and Venkatasubramanian, K., Enzyme Engineering (Vol. 4), G. Broun and G. Manecke (editors), Plenum Press, New York (in press).

6. Vieth, W.R., Gilbert, S.G., Wang, S.S., and Venkatasubramanian, K., U.S. Patent 3,809,613 (1974).

RECEIVED February 15, 1979.

Pore Dimensions for Accumulating Biomass

R. A. MESSING, R. A. OPPERMANN, and F. B. KOLOT

Corning Glass Works, Sullivan Park, Corning, NY 14830

Such processes as the production of cells (single cell proteins), fermentation for secondary metabolites, and waste conversion require high concentrations of cells. Generally speaking, to produce a large number of cells such as is required for single cell proteins, one must pass through the lag phase and operate within the log phase. The greater the number of cells per unit volume, the more progeny will be produced per volume provided that the cells are neither nutrient limited nor gas limited.

The requirement for high concentration of cells or accumulations of biomass in the production of secondary metabolites is even more apparent than for cell production. Secondary metabolites are generally produced in the stationary phase; the greater the concentrations of cells, the greater the production of secondary metabolites per unit volume and per unit time.

The accumulation and retention of biomass lends itself readily to the employment of continuous single pass reactors such as plug-flow or fluidized-bed reactors. When high quantities of biomass are retained, greater total quantities of nutrients may be delivered and greater quantities of waste products may be removed per unit time. Any mechanism that can be offered to retain the cells in high concentrations, deliver the nutrients rapidly, and remove waste products should offer a highly efficient reactor.

We have found a relationship between the accumulation of stable and viable biomass and the pore morphology of a dimensionally stable inorganic carrier. That relationship is dependent upon the mode of reproduction of the specific microbe.

0-8412-0508-6/79/47-106-013$05.00/0

Materials and Methods.

The controlled-pore ceramics and the fritted
glasses employed in these studies were manufactured by
Corning Glass Works. The borosilicate glass is a
product of Corning Glass Works.

A DuPont Biometer, Model 760, was employed to
determine the viable microbe count for all organisms
except Streptomyces olivochromogenes, and Penicillium
chrysogenum. We employed a protein determination
utilizing the Folin reagent for determining the
loadings of the latter two microbes.

Determination of Microbial Loadings(Biomass).

We were not able to employ the conventional plate
counting techniques to determine loadings due to the
fact that microbial loadings (biomass accumulations)
involved measuring the number of microbes bonded
within the pores of various porous supports. Instead,
the microbe counts or the relative quantity of biomass were determined by employing the Biometer which
determines the relative number of viable microbes or
viable biomass based upon the amount of ATP present
in a given sample. The actual procedure employed was
as follows: to approximately 10-20 mg of immobilized
microbe composite, 0.5 ml of 90% dimethylsulfoxide in
water was added and the suspension was mixed vigor-
ously for 10 seconds. The suspension was allowed to
extract for 20 minutes and then 4 ml of 0.01M
morpholinopropane sulfonic acid buffer, pH 7.4, was
delivered and the suspension was mixed thoroughly and
stored in ice until preparations were complete for
the ATP determinations. Prior to this determination,
the luciferin-luciferase mixture was prepared
according to the Biometer procedure[1]. After 0.1 ml
of enzyme-substrate mixture was delivered to the
reaction cuvette, 10 microliters of the above DMSO
extract was added to the cuvette containing the luciferin-luciferase mixture. The light emission measurement was then made and the correlated value to viable
cell quantity was recorded. Since it is impossible to
relate individual cells of microbes that generate
mycelia to ATP present, we elected to report the mass
of ATP in femtograms (10^{-15} gms) as representative of

the viable biomass with respect to loading/gram for
each individual carrier.

   The determination of biomass protein for both S.
olivochromogenes and P. chrysogenum was performed on
0.5 gms of carrier which contained the immobilized
biomass. After the loadings and/or subsequent growth,
the 0.5 gm of carrier sample was washed 3 times with
10 ml of phosphate buffer, pH 7.0. The carrier was
then extracted with 3 ml of phosphate buffer by
grinding therewith. Subsequently, 3 ml of 1 N sodium
hydroxide were added, the temperature of the mixture
was raised to 60° and maintained at that temperature
for 1 hour in order to hydrolyze the mycelia. This
procedure results in quantatative release of the
protein from the mycelia.

   To further extract protein, 3 ml of ethyl alcohol
were added to the mixture in the case of S. olivochromogenes while 2 ml of the alcohol were added in the
case of P. chrysogenum and this was allowed to react
at room temperature for one-half hour. After centrifugation, the supernatant fluid was decanted and the
protein content of the fluid was determined according
to the procedure of Hill et al[2] and Hauschka[3].

   Table I

Carrier Parameters

Carrier Number	Average Pore Diameter(μ)	Pore Diameter(μ) Distribution	Carrier Composition
1	1.1	0.8-1.8	Fritted glass
2	3.0	1.5-6	Cordierite
3	3.1	1.5-4	Fritted glass
4	3.5	1.5-4.5	Fritted glass
5	4.5	3-6	Fritted glass
6	10	2-19	Cordierite
7	13	8-20	Fritted glass
8	19	17-35	Zirconia Ceramic
9	40	18-100	Fritted glass
10	195	170-220	Fritted glass
11	non-porous	____	Borosilicate glass

Bioaccumulation of Escherichia coli.

   Samples of E. coli (having major dimensions of
1-6μ) were immobilized to the various carriers, using
sterile technique on both a virgin carrier surface and
a surface that had been converted to a terminal amine
by silanization. The bonding by adsorption to the
virgin surface was accomplished by reacting 3 gm portions of 18-25 mesh particles of the indicated carriers
for 3 hours at 22°C with 20 ml of suspension of E.
coli cells. The bonding to the terminal amine surface
was initiated first by reacting 2 gms of 18-25 mesh
particles with 20 ml of 10% γ-aminopropyltriethoxy-
silane in water for 2 hours at 100°C. The treated
carrier was subsequently dried and reacted with 20 ml
of suspension of E. coli cells overnight at 22°C.
Bacterial loadings (number/gram of support) were then
recorded via ATP measurement approximately 18 hours
after the preparation of the immobilized microbes.
The results are summarized in Figures 1a and 1b.

Bioaccumulation of Serratia marcescens by Polyisocyanate Coupling.

   S. marcescens, having major dimensions of 0.6-2μ
was coupled to glass surfaces with polyisocyanate
(PAPI 901, Upjohn Company, Kalamazoo, MI). The
carrier derivative was prepared by shaking at 100 RPM
0.5 gms of carrier in 10 ml of 0.5% polyisocyanate in
acetone for 45 minutes at room temperature. The
coupling solution was decanted and replaced with 10 ml
of a cell suspension containing 3 x 10^9 cells/ml. The
cells were reacted with the derivatized carrier for 3
hours following which the excess cells were poured off
and the carrier was washed 3 times with 0.1 M phosphate
buffer at pH 7.2. The results of this study with
various carriers are plotted in Figure 2.

The Immobilization of Bacillus subtilis by Adsorption.

   The culture cells were sized prior to immobilization. The major dimension of the cells were found
to be between 3 and 4μ with some long double cells of
approximately 7μ in length.

Prior to use, the carriers were dry autoclaved.

Bacillus subtilis cells were grown in 1 liter brain heart infusion broth for 36 hours, centrifuged,and the cells were washed 2 times with sterile phosphate buffer Ten ml of the washed cell suspension was added to each flask which contained 0.5 gms of carrier. After 3 hours of contact time, the carriers were washed 3 times and maintained at 8°C overnight.

The cell mass immobilized on the carriers was determined by the Biometer. The results of this study are reported in Figure 3.

Bioaccumulation of Yeast.

The yeast cells employed in this study were all grown in shake flasks in nutrient broth plus 1% dextrose for a period of 36-40 hours at room temperature. The cell suspension was centrifuged and washed 3 times with sodium-potassium phosphate buffer, pH 7.2. The washed suspension of cells was added to the carrier and agitated by shaking for 3 hours of contact time.

For adsorption, the carriers were ground to an 18-25 mesh, sterilized dry, and placed in a 37°C incubator overnight to produce a dry carrier. A quantity of 0.5 gms of each carrier was added to 50 ml flasks. Ten ml of concentrated yeast cell suspension was added to the carrier in the flask. At the end of 3 hours of contact time, the excess cells were poured off, the carrier was washed 3 times with phosphate buffer and stored in the refrigerator prior to the loading determinations.

The coupling procedure involved the addition of 0.5 gms of each carrier to separate 50 ml flasks. The flasks, with contents, were dry autoclaved and placed in a 37°C incubator overnight to maintain the carrier dry. Ten ml of a 0.5% polyisocyanate in acetone solution was added to each carrier. The flasks with contents were shaken for 45 minutes at room temperature after which the coupling solution was decanted. Ten ml of concentrated yeast cell suspension was then added to each flask and the flasks were shaken for 3 hours after which the excess cells were decanted and the immobilized preparation was washed 3 times with phosphate buffer.

The ATP determinations of cell loadings were performed with the Biometer as previously described.

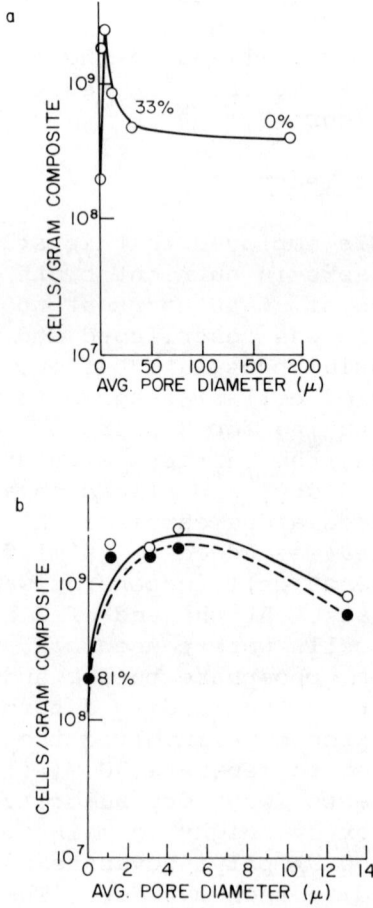

Figure 1. (a and b) Bioaccumulation of E. coli

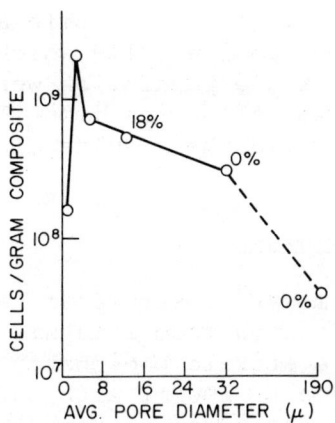

Figure 2. Bioaccumulation of S. marcescens

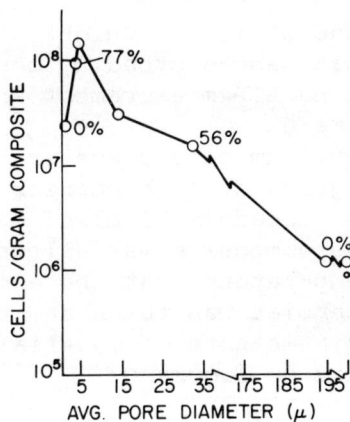

Figure 3. Bioaccumulation of B. subtilis

The yeast, <u>Saccharomyces cerevisiae</u>, was sized prior to immobilization. The average cell dimensions were 4 x 5.5μ (2.5-4 x 4-7) and about 20% of the cells were 4.5 x 7μ. The results of the bioaccumulation of this microbe by adsorption are plotted in Figure 4.

The yeast, <u>Saccharomyces amurcae</u>, prior to immobilization were found to have dimensions of 5.5 x 7μ (3-7 x 6-9) in terms of small, single cells while an additional population of larger, double cells which constituted approximately 75% of the population were found to be 6-8 x 13-18μ. These cells were immobilized by coupling with polyisocyanate and the results of this study are plotted in Figure 5.

Bioaccumulation of Aspergillus niger.

The spores of this organism were observed to range from 3-5μ. These spores were eluted from a mature growth in a Blake bottle with a sterile phosphate buffer. The suspension was made to 100 ml with buffer. Ten ml of the spore suspension was added to each 50 ml microfernback flask which had been dry autoclaved with 1 gram of carrier and dried overnight prior to its use. After 3 hours of shaking at room temperature, the excess spores were poured off, the carrier was washed 3 times with phosphate buffer and stored overnight at 8°C. The quantity of spores that was adsorbed by the various carriers was determined by ATP measurement and the results are plotted in Figure 6.

In order to determine the optimum pore diameter range for mycelial growth, 0.5 grams of each carrier with the immobilized spores was placed in 75 ml of Sabouraud dextrose broth and the composite was allowed to shake on a shaker at room temperature. At the end of 27 hours, a sample of each carrier was taken and the amount of ATP was determined as a measure of mycelial growth. The results are recorded in Figure 6.

Bioaccumulation of <u>Streptomyces olivochromogenes</u>.

Equal amounts of spores in phosphate buffer were added to 0.5 grams of each of the carriers. The spores and carriers were allowed to react together for 48 hours. The non-reacted spores were decanted and the carrier was washed 3 times with 2 ml aliquots of

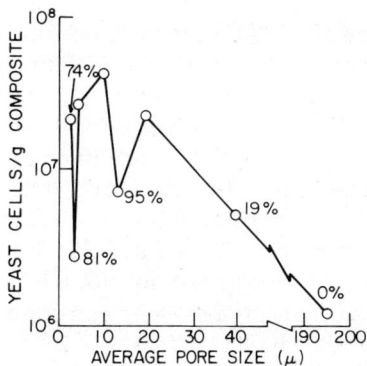

Figure 4. Immobilization of S. cerevisae *into various carriers by absorption*

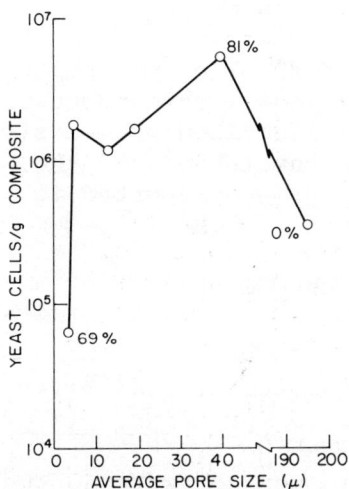

Figure 5. Immobilization of S. amurcae *into various carriers by chemical coupling*

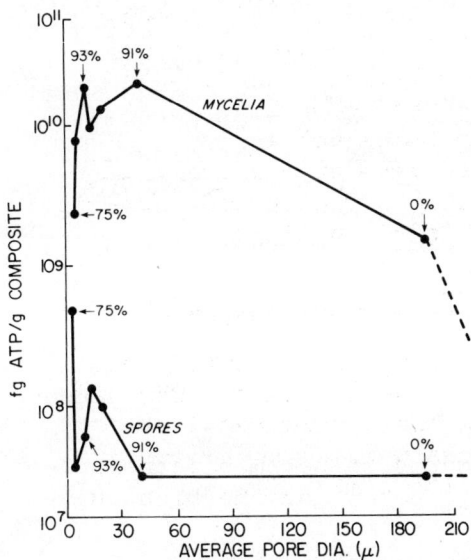

Figure 6. Bioaccumulation of A. niger *in controlled pore inorganics*

sterile phosphate buffer. These non-reacted spores
plus the washings were collected, the volume determined
and then analyzed for protein content. The quantity
of protein adsorbed in the carrier was calculated as
the difference between the initial content of protein
in the reacting volume containing the spores and that
contained in the non-reacted spores plus the washings.

The washed carrier containing the spores was
separately transferred to a flask containing 50 ml of
Emerson broth medium. Mycelium formation within the
carrier pores was evaluated after incubation with
shaking for 24 and 48 hours at room temperature.

At the end of the stated periods of incubation,
the carrier was separated from the medium via centrifugation and washed 3 times with 10 ml of phosphate
buffer, pH 7.0. The protein content within the carrier
was determined by the procedure described previously.
The results of both the spore accumulation and the
mycelial growth within the carrier are reported in
Figure 7.

Bioaccumulation of Penicillium chrysogenum.

The techniques for the immobilization of the spores
and the handling of the carrier materials was essentially the same as those described for the S. olivochromogenes experiments. The washing of the carriers
and the determination of protein content were conducted
in the same manner as previously described. Other than
the modification of alcohol quantity for extracting the
protein, the only other change was the medium for
mycelial growth. In place of the Emerson broth, 50 ml
of an aqueous medium having a pH of 6.3 with the
following composition was utilized for the growth of
P. chrysogenum: 2% lactose, 1% glucose, 0.2%KH_2PO_4,
0.125% NH_4NO_3, 0.05% Na_2SO_4, 0.025%$MgSO_4$, 0.002% $MnSO_4$,
0.00025% $CuSO_4$, and 0.002% $ZnSO_4$. A single incubation
of 48 hours was employed.

The results obtained with P. chrysogenum are
recorded in Figure 8.

Results and Discussions

Upon perusal of the figures,it becomes rather clear
that at least one optimum is noted for the bioaccumu-

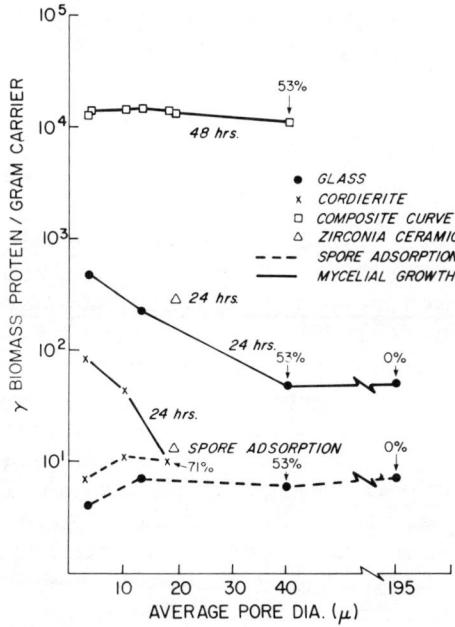

Figure 7.   *Bioaccumulation of S. olivochromogenes in controlled pore inorganics*

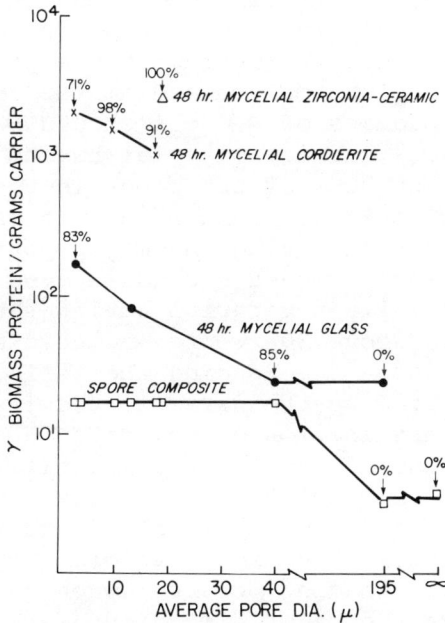

Figure 8.   *Bioaccumulation of P. chrysogenum in controlled pore inorganics*

lation of each microbe when one plots the number of
viable cells or biomass/gram versus the average pore
diameter of the carrier. If we now attempt to define
the critical parameter of the microbe which governs
the optimum loading, it becomes necessary to divide
these studies into three groups. The first group of
studies include those microbes that reproduce by
fission (E. coli, S. marcescens, B. subtilus), the
second group consists of those that reproduce by
budding (Saccharomyces cerevisiae, Saccharomyces
amurcae) and the third group consists of those microbes
that exhibit mycelial growth and produce spores (A.
niger, S. olivochromogenes, P. chrysogenum).

In the case of microbes that reproduce by fission,
the maximum accumulation of stable and viable biomass
occurs when the pore diameter is between 1 and 5 times
the major dimension of the microbe. Since it is
exceedingly difficult to obtain a porous material that
contains pores only of one diameter, and since there
is a considerable variation in the major dimension of
a microbe, it is very difficult to further refine this
relationship. The variability of the length of a
microbe may best be illustrated in the case of E. coli
whose major dimension may vary in a single culture
from 1 to 6 microns. According to the previous hypothesis, then, the optimum pore diameter for accumulating a high biomass of E. coli should be between 1
and 30μ. Thus, a carrier material having a major
portion (at least 70%) of the pores in the 1-30μ
diameter range would accumulate the highest biomass of
E. coli. A material having the major portions of its
pores under 1 micron would exclude a good percentage
of these cells while a material having the predominant
number of pores above 30μ would not have as high a
usable surface area for bonding the cells. This, in
fact, is supported by the data plotted in Figures 1a
and 1b. Upon examination of the data in Figure 2 for
S. marcescens and Figure 3 for B. subtilus, it may be
noted that the same relationship holds.

A previous study[4] which involved the relationship
of enzyme immobilization and pore diameter indicated
that the optimum loading occurred when the pore diameter was twice the major dimension of the enzyme.
Logically speaking, this should have been the same
case for microbes; however, if we consider not only the

variation of the microbe size but, in addition, the
fact that the microbe increases in size as it undergoes
reproduction by fission, then it is clear that the
optimum pore diameter would of necessity be substantially greater than twice the major dimension of the
cell. In order to examine this situation, let us consider a model system in which all of the cells have
the major dimension of one micron. When the cell
undergoes fission, it increases in length to approximately 2 microns. If two of these cells just prior to
separating were immobilized opposite each other on the
walls of the pores, in order to undergo fission there
must be at least 4μ in pore diameter. Now if another
cell required passage beyond those two dividing cells,
at least another micron would be required for a total
of 5 microns. Therefore, a minimum of 5 microns or 5
times the major dimension of the microbe would be required for optimum growth and accumulation of cells
within the pore. In fact, in the three studies of
microbes that reproduce by fission, it may be noted
that 5 times the smallest figure given for the major
dimension appears to be the peak value for biomass
accumulation independant of the mode of immobilization.

In the second case, that of reproduction by
budding exemplified by the yeast studies in Figure 4
and 5, the rationale for the dimensions of optimum
pore can be expressed in the following manner. When a
cell is immobilized on the wall of a pore that is
opposite another immobilized cell, and each· cell produces a bud proximally to the opposite cell, their
length will increase by approximately 1¼-1½ times.
This produces cell masses on opposite sides of the
channel requiring pores approximately 3 times the
cell's total length. Now, if a third cell is to pass
the 2 fixed on the walls, a proper channel must be
left which would be approximately 1 major cell
dimension. Therefore, to allow passage and immobilization of yeast, we must make use of pores of
approximately 4 times the maximum dimension of the
yeast cell employed.

In each of the figures, percentages appear at those
points plotted on the curve. This figure represents
the percent of the pores in the carrier that fall
within the range of the smallest to the largest pore
diameter for the optimum bioaccumulation of that

particular microbe. Those points that do not contain
a percentage figure have 100% of their pores falling
within the optimum range.

It can be noted in the graph (Figure 4) for
Saccharomyces cerevisiae that a single, fairly broad
curve extending from 3 to 40μ contains two peaks.
This observation correlates very well with the fact
that the average cell dimensions were 4 x 5.5μ (2.5-4
x 4-7) while 20% of the cells were 4.5 x 7μ in length.
The amount of variability within this culture accounts
for the broad, double peak. The sharp drop at 3.5μ
was not caused by the size of the average pore but
rather by the reduction in the large pore diameters
which appear to optimally accumulate as indicated by
the data for the 4.5 average pore diameter carrier.

Saccharomyces amurcae was studied by immobilization
on various carriers by chemical coupling. In this
study, because the culture was basically biphasic,
small single cells and large duplicating cells, the
graph (Figure 5) again shows a double peak with the
major peak in the range of 13-40μ. This correlates
well with the observation that the average small cells
were 5.5 x 7μ (3-7 x 6-9) and that the larger, double
cells which constituted approximately 75% of the population were 6-8 x 13-18μ long.

The yeast experiments demonstrate that independent
of the method of attachment, the optimum pore diameter
for accumulating biomass of a budding form is 1 times
the smaller dimension to 3-4 times the largest of that
microbe.

The results recorded in Figure 6 for the bioaccumulation of A. niger indicate that although the highest
quantity of spores was immobilized in the smallest pore
diameter material (3.5μ), the pore was not large enough
to allow a great deal of mycelial growth. The average
size of these spores was 4μ (3-5μ). A key point to note
is that the highest recorded value for mycelial growth
was in a material that had an average pore diameter of
40μ and only 91% of those pores were in the range that
we believe is optimum for the bioaccumulation which is
1 times the smallest diameter (3μ), and 16 times the
largest spore diameter (5μ). Although the spore immobilization curve indicates that approximately the same
number of spores was immobilized in the 40μ material as
was immobilized on the non-porous glass, a marked in-

crease in mycelial growth was noted with the 40μ glass while neither the 195μ or the non-porous glass demonstrated the same level of growth.

Since an uninterrupted plot between the 40μ and the 195μ glass in the growth curve indicates that 80μ average pore diameter would allow growth equivalent to that noted at 13μ, it appears clear that the upper limit of pore size is 80μ which is 16 times the largest spore diameter (5μ).

The information in Figure 7 for S. olivochromogenes which exhibits spores of between 1-2.5μ in diameter indicates a marked surface effect on spore adsorption and mycelial growth by the various inorganic carriers. The more negative fritted glass adsorbs fewer spores but allows much more mycelial growth in the first 24 hour period than the less negative surface of the cordierite-type crystalline material.

Based on the data recorded for a single carrier, it appears that zirconia ceramic adsorbs more spores than either the cordierite or the glass. It also should be noted, however, that this surface effect is not permanent but rather is substantially eliminated after 48 hours of growth. This phenomenon is not totally unexpected and can be explained in the following manner. The first deposition or monolayer of cells will obviously be effected via direct contact with the carrier surface. Subsequently as growth continues and this new growth becomes more and more remote from the surface, the dimensions of the carrier pores exert a much greater effect on bioaccumulation than does the surface.

The diameter of S. olivochromogenes spores range between 1 and 2.5μ. An examination of the curve representing mycelial growth on fritted glass after 24 hours and the curve representing the composite of mycelial growth on the carriers after 48 hours indicates that a high biomass accumulation occurs at about 16 times the largest dimension of the spores (16 times 2.5μ = 40μ). It should also be noted that in the fritted glass having an average pore size of 40μ, only 53% of the pores were between 1-40μ rather than the 70% suggested to obtain optimum accumulation. Under these circumstances, the accumulation of biomass was not significantly different from that obtained with fritted glass having an average pore diameter of

195μ where none of the pores fall within the 1-40μ range.

Our studies of P. chrysogenum, Figure 8, whose spores range between 2.5-4.5μ in diameter, indicate that contrary to the behavior noted with respect to S. olivochromogenes, little or no surface contribution on spore bioaccumulation occurred. After mycelial growth of 48 hours, however, the P. chrysogenum appeared to prefer the less negative cordierite over the very negatively charged fritted glass. This, of course, is diametrically opposed to the results with S. olivochromogenes.

Although the mycelial growth curve on glass appears to have leveled off at a pore diameter of 40μ, only 85% of those pores are within the range of 2.5μ (1 times the smallest dimension) and 72μ (16 times the largest dimension). Since the spore adsorption curve exhibits a tremendous drop between 40μ and 195μ and the 48 hour mycelial growth curve for cordierite appears to parallel the 48 hour mycelial glass curve but is significantly higher, it would appear as though major accumulation would still occur at 72μ which is 16 times the largest spore dimension.

Literature Cited.

1. Instruction Manual DuPont 760 Luminescence Biometer, 760 M3. December 1970, pgs. 3-2, 3-3.
2. Hill, E. C., Davies, I., Pritchard, J. A., Bryon, D., Journal of the Institute of Petroleum 53, 275 (1967).
3. Hauschka, P. V., Analytical Biochemistry 80, 212 (1977).
4. Messing, R. A., Biotechnol. Bioeng. 16, 897 (1974).

RECEIVED February 15, 1979.

3

The Biophysics of Cellular Adhesion

D. F. GERSON and J. E. ZAJIC

Faculty of Engineering Science, The University of Western Ontario,
London, Ontario, Canada

Immobilization of whole microbial cells for industrial purposes eliminates the need for the isolation, purification and
attachment of enzymes and provides the enzymes with a microenvironment maintained at optimal conditions by cellular metabolic
and transport activities. Adhesion of microbial cells to inert
substrata often occurs in nature and greater understanding of
these natural processes may lead to advances in the technology of
whole cell immobilization. Mechanisms of attachment in natural
systems involve adhesive microexudates produced by the cells,
electrostatic attraction, and anatomical projections which cling
to the support surface. The chemical methods which have been
used for whole cell immobilization have recently been reviewed
by Jack and Zajic (1).

I. The Surface Physics of Adhesion

A. Surface Free Energies. Surface free energies must
dominate any explanation of the adhesion between different phases
which are not mechanically linked. Current levels of understanding of adhesiveness are such that actual adhesive strengths
are always much less (1-0.1%) than those predicted by thermodynamic analysis, and often there is apparently little correlation between the two. Further refinement of the theory of
adhesiveness will require understanding of the importance of
flaws in an adhesive joint and of the relative contributions of
polar and dispersive Van der Waal's interactions. The following
is an analysis of adhesion in terms of surface free energies.
The surface of a substance or phase requires an extra term
in the description of its energy which is due to its location
at a boundary where there is a sharp change in concentration or
properties of the substance. The energy associated with this
location is the surface free energy and is described by the
energy required to form a new unit area of surface (Eq. 1).

0-8412-0508-6/79/47-106-029$07.00/0
© 1979 American Chemical Society

$$\frac{dG}{dA} = \gamma \qquad\qquad (1)$$

where: G = Gibbs free energy
         A = area
         γ = surface free energy per unit area

Surface tension (the force required to part a unit length of surface) is equivalent to surface free energy for liquids, but this
is not necessarily true for solids.
    Surface energies for some liquids and solids are given in
Table 1. It is relatively simple to measure these values for
liquids by drawing a wettable solid through the interface and
measuring the force required to part a unit length of the interface. Clearly, this is not easily done for a solid. (Breaking
strength, see below, depends on too many practical considerations
to be an accurate measure of solid surface tension). The most
successful method for measuring solid surface energies is due to
Zisman (2,3). In this method (Fig. 1), the surface energy of the
solid with respect to a particular liquid (or liquid type) is
given by the maximum liquid surface tension required for wetting
the solid surface (zero contact angle). This is the critical
surface tension, γ_c, and is a measure of the solid surface energy, γ_s, but may not be identical to it. In general, γ_c is less
than γ_s, and this error results from a variety of interactions
between the particular solid and liquid under study (see below).
The method is also applicable to the study of liquid surfaces.

    B. Solid Surface Free Energies. The determination of the
surface energy of solids (γ_s) involved in an adhesive joint is
crucial to any calculations of the thermodynamic properties of
that interface. Zisman's method involves determination of the
contact angles of a series of similar liquids (e.g. n-alkanes)
having a range of surface tensions. The usual result is that as
liquid surface tension decreases, the cosine of the contact angle
on a given solid increases linearly until spreading occurs ($\theta=0°$,
$\cos\theta=1$). The maximum liquid surface tension which will result in
spreading is the critical surface tension γ_c. This empirical
result is given by Equation 2.

$$\cos\theta = 1 + b(\gamma_c - \gamma_{\ell v}) \qquad\qquad (2)$$

where: θ = contact angle of the liquid on the solid
         $\gamma_{\ell v}$ = liquid-vapour surface energy
         b   = slope of line relating $\gamma_{\ell v}$ to $\cos\theta$

This approach to the determination of γ_s is generally useful, but
suffers from observed non-linearities when polar liquids are used.
More thorough understanding of the relations between measured
contact angles and solid surface tensions depends on the combina-

Table I

Surface Tensions of Liquids and Surface Energies of Solids

a) Liquids

Substance	Surface Tension (dynes/cm)
NaNO$_3$, 30% (w/v), aqueous	80.5
CaCl$_2$, 10% (w/v), aqueous	74.9
water	73.0
glycerol: 20% aqueous	70.9
82% aqueous	65.3
pure	63.4
ethyleneglycol, 55% aqueous	55.6
n-hexadecane	27.3
polyethylene glycol-6000	24
ethanol	22.3
n-hexane	18.4

b) Solids

Substance	Surface Energy (ergs/cm^2)
platinum	≈1800
glass	≈ 700
mica	≈ 350
ice	106.0
poly(hexamethyleneadipamide)[Nylon-66]	46.0
glass, 0.6% relative humidity	45.0
poly(ethyleneterephthalate)[Mylar]	43.0
poly(vinylidene chloride)	40.0
poly(vinyl chloride)	39.0
Tellurium	35.5
poly(styrene)	33
Selenium	32
poly(ethylene)	31.0
glass, 95% relative humidity	≈31.0
Sulfur: amorphous	31.5
monoclinic	30.5
orthorhombic	30.0
poly(vinyl fluoride)	28.0
poly(vinylidine fluoride)	25.0
n-hexatriacontane	21.0
poly(tetrafluoroethylene)	18.5
poly(perfluorostyrene)	17.8
poly(perfluoropropylethylene)	15.5
poly(1H,1H-pentadecafluorooctylmethacrylate)	10.6

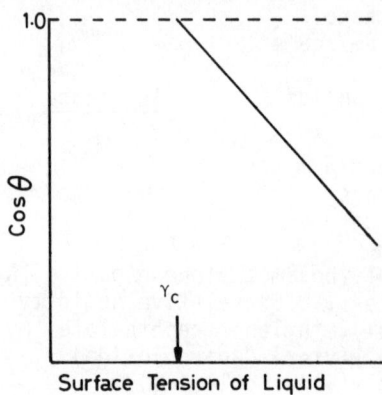

Figure 1. Plot of liquid surface tension vs. cosine of contact angle on a hypothetical solid. Spreading occurs when cos θ = 1 and the corresponding surface tension, γ_c, is the critical surface tension of the solid.

tion of a basic analysis of the forces maintaining drop shape on
a surface and the experimental correlations between contact angle
and liquid surface tension. Young's equation (Young, 4) describes the balance of forces present for a liquid drop located
on a surface (Eq. 3). The contact angle, θ, is the equilibrium

$$\gamma_{sv} = \gamma_{s\ell} + \gamma_{\ell v} \cos\theta \qquad (3)$$

where: γ_{sv} = solid-vapor interfacial free energy

$\gamma_{s\ell}$ = solid-liquid interfacial free energy

contact angle, which in practice is difficult to determine since
contact angle shows hysteresis between an advancing maximum and
a receding minimum. In application to a particular problem,
however, one of these is usually more relevant than the other;
their average may be close to the equilibrium value. Experimentally, a strong correlation was found between γ_{sv}, calculated
from Equations (2) and (3) ($\gamma_c \approx \gamma_{sv}$), and the interaction parameter, θ, given by Equation 4 (Good and Girifalco, 5 and Good,
6). The correlation is given by Equation 5.

$$\phi = \frac{\gamma_{sv} + \gamma_{\ell v} - \gamma_{s\ell}}{2\sqrt{\gamma_{sv}\gamma_{\ell v}}} \qquad (4)$$

$$\phi = -.0075 \, \gamma_{s\ell} + 1.00 \qquad (5)$$

Use of the interaction parameter to correct for non-linearities
present in plots of $\cos\theta$ vs $\gamma_{\ell v}$ resulted in the following equations (Eq. 6,7)(Neumann et al, 7,8) which represent a great
advance in bringing surface chemistry within experimental reach,
and allow calculation of solid surface energies from easily
measured contact angles and liquid surface energies. Equation 6
is cubic in $\gamma_{\ell v}$ and some care must be taken in selecting the

$$\cos\theta = \frac{\gamma_{\ell v} + (0.015\gamma_{sv} - 2.00)\sqrt{\gamma_{sv}\gamma_{\ell v}}}{\gamma_{\ell v}(0.015\sqrt{\gamma_{sv}\gamma_{\ell v}} - 1)} \qquad (6)$$

$$\gamma_{s\ell} = \frac{(\sqrt{\gamma_{sv}} - \sqrt{\gamma_{\ell v}})^2}{1 - 0.015\sqrt{\gamma_{sv}\gamma_{\ell v}}} \qquad (7)$$

correct root. Neumann et al (8) have published a Fortran pro-
gram which incorporates selection criteria for the most appropriate value of γ_{sv}.

There is always some dependence of measured values of γ_c on

the liquid used to make the measurement. These result from the
net combination of dispersive, polar and other interactions between the two phases and there have been a variety of attempts to
explain, predict or cope with this problem in order to better
predict γ_s from γ_c. Good (6) has approached this through use of
the interaction parameter ($\overline{Eq}$. 4), and has shown that the recip-
rocal of ϕ is the coefficient relating these when $\theta=0°$ (Eq. 5a).

$$\gamma_s = \frac{\gamma_c}{\phi^2} \tag{5a}$$

If the data are available, a good estimate of ϕ may be obtained
from values of the polarizability, dipole moment and ionization
energies of both the liquid and solid phases (Good, 9, Good and
Elbing, 10). Kitazaki and Hata (11) assumed $\gamma_{s\ell}$ was non-zero at
γ_c and were able to separate values of γ_c due to non-polar liquids (e.g. n-alkanes), polar liquids (e.g. chloroform) and hydrogen bonding liquids such as water or glycerol. Rhee (12) combined Zisman's empirical equation with Young's equation to obtain
a relation between γ_c and γ_s, but this of course includes the
known inaccuracies of Zisman's empirical relation (Eq. 2).

   C. Free Energy of Adhesion. The thermodynamic work of
adhesion, W_a, describes the energy of separation at an interface
and simply is the difference between the surface free energies of
newly formed interfaces and the surface free energy of the interface prior to separation. For a liquid-solid joint separated in
a vacuum, the free energy of adhesion is given by Equation 8a;
for separation in an atmosphere of the liquid vapor it is given
by Equation 8b. These free energies of adhesion are related by
the spreading pressure or the energy of adsorption of the vapor

$$W_a = \gamma_s + \gamma_\ell - \gamma_{s\ell} \tag{8a}$$

$$W'_a = \gamma_{sv} + \gamma_{\ell v} - \gamma_{s\ell} \tag{8b}$$

on the solid, π_e, however this is usually considered to be neg-
ligible (Mittal, 13). In the following, it is assumed that
$W_a = W'_a$. Equation 8 is generally applicable; but, for low
energy solids and liquids which display a finite contact angle,
Equation 9 may be used to calculate the free energy of cohesion
(from Eq. 8 and Eq. 3, i.e. the Young-Dupré equation).

$$W_a = -\gamma_{\ell v}(1 + \cos\theta) = -\Delta G_a \tag{9}$$

   where: ΔG_a = free energy of adhesion

   D. Adhesive Strength. Even though calculated free energies of adhesion are 100-1000 times larger than the observed strengths of adhesive joints, there are some generalizations which describe the conditions for optimal adhesive strength. One criterion which is well borne out in practice is the desirability of a zero contact angle between the adhesive and the adherand solids. In terms of microbial immobilization, measurements of the contact angle of the glycocalyx or microexudate material on potential substrates for immobilization could be an extremely useful pre-screening procedure. Given the desirability of a zero contact angle (e.g. $\gamma_{\ell v} < \gamma_{sv}$) several different approaches have yielded the same criteria for maximum adhesion (Kitazaki and Hata, 11 and Mittal, 13). From Zisman's empirical equation (Eq. 2), the equation for adhesion tension (Eq. 10) and the assumption that

$$\tau = \gamma_{\ell v} \cos\theta \tag{10}$$

where: τ = adhesion tension

maximum adhesion tension corresponds to maximum adhesive strength, one can derive the following relation between $\gamma_{\ell v}$ and γ_c for maximum adhesiveness (Eq. 11). This relation produces fairly reasonable estimates of $\gamma_{\ell v}$ for a given γ_c and the residual error

$$\gamma_{\ell v} = \frac{1}{2b} + \frac{\gamma_c}{2} \tag{11}$$

may be associated with the errors implicit in Equation 2. In addition, Equation 11 corresponds to the condition of maximum penetration into cracks and crevices in the solid materials, thus minimizing voids and other faults in the adhesive bond.

Another criterion for maximum adhesive strength (Gray, 14) is a minimum $\gamma_{s\ell}$. This is especially interesting in terms of the work of Albertsson (15) which demonstrates that the interfacial tensions between mutually immiscible colloidal phases can be extremely low (e.g. 0.001 dynes/cm). If this is the case for the interfacial tension between the glycocalyx and the microbial cell wall, then this part of the adhesive joint could be relatively strong. A similar criterion could be applied to selection of immobilization substrates.

In a good adhesive bond, the joint is able to transmit both tensional and shear forces across the interface. Generally, an adhesive bond will be weaker than either bulk material joined by it since the predominant forces at the interface are Van der Waals while those in a bulk material are usually covalent. If one phase is a microorganism, this may not be true since a proportion of the "bonds" holding the organism together are not covalent (e.g. hydrophobic association in the membrane; glycolipid-protein complexes). In macroscopic adhesive bonds, flaws are the primary source of weakness; since the area for bacterial adhesion

to a solid is extremely small and since the bacterium itself is
most commonly the source of adhesive materials, it may be possible that cells will form joints which are very nearly free of
flaws, and thus are more closely comparable in strength to the
thermodynamic adhesion energy.

E. Free Energy of Entrapment. One excellent method for
immobilization of whole cells is entrapment in droplets of an
immiscible liquid or polymer (Jack and Zajic, 1 and Larsson et
al, 16). The surface physics of this situation are quite similar to those described by Neumann et al (17) for the engulfment
of particles by a solidifying melt (or polymerizing polymeric
bead). The stages of this process are diagrammed in Fig. 2, and
the free energy changes are as follows: The overall free energy
change is simply the difference between the interfacial free
energies prior to and following entrapment (Eq. 12).

$$\Delta Gnet = \gamma_{c\ell} - \gamma_{cp} \tag{12}$$

where: $\Delta Gnet$ = net free energy change
         $\gamma_{c\ell}$ = cell-liquid interfacial free energy
         γ_{cp} = cell-polymer interfacial free energy

If it is assumed that a microbial cell is spherical, then
the entrapment process proceeds by 2 steps: partial immersion
(step 1, Fig. 3) and total entrapment (step 2, Fig. 3). If ΔG_1
is negative, attachment is spontaneous, and if ΔG_2 is negative,
complete engulfment by the polymeric droplet is spontaneous (ΔG_1
could be positive while ΔG_2 is negative, but the consequences of
this are uncertain). Calculation of these free energy changes
utilizes the equations described above and requires knowledge of
the relevant surface and interfacial free energies. The surface
energy of living cells has been measured by van Oss et al (18)
using contact angles and relatively simple techniques, so this
does not pose a significant problem (see below). It should be
noted, however, that cells of different species and even cells of
different strains of the same species may have widely different
surface free energies, and thus the surface energy of the cells
should be measured under conditions close to those which will
exist at the time entrapment or attachment is attempted. It is,
of course, possible to increase entrapment efficiency by rapid
solidification of the entrapping polymer and overcoming the implications of equilibrium thermodynamics by irreversible conditions. The quality of entrapment or the strength of attachments
made under these conditions is not predictable.

F. Molecular Properties and Surface Free Energies. Ultimately, both adhesion phenomena and surface free energies are a
consequence of the molecular properties of materials. These

Table II

Electrodynamic Attractive Forces and Work of Adhesion Between Cells and Solid Surfaces

Solid	Electrodynamic Attractive Force $(dynes/cm^2)$*	W_a $(ergs/cm^2)$**
Platinum	9.0×10^4	≈ 1800
Quartz	2.4×10^4	915
Polystyrene	1.8×10^4	79
Polyethylene	1.8×10^4	76
Teflon	0.7×10^4	54

*Parsegian and Gingell, (19)
**γ_{sv} of cell taken as 68 $ergs/cm^2$

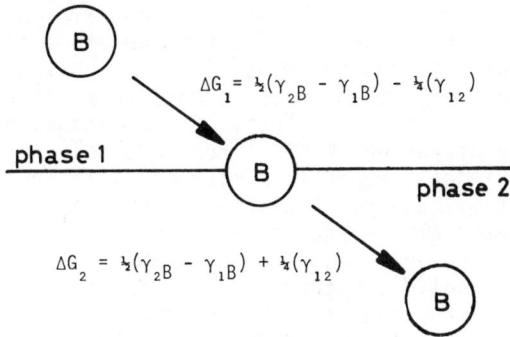

$$\Delta G_1 = \tfrac{1}{2}(\gamma_{2B} - \gamma_{1B}) - \tfrac{1}{4}(\gamma_{12})$$

phase 1

phase 2

$$\Delta G_2 = \tfrac{1}{2}(\gamma_{2B} - \gamma_{1B}) + \tfrac{1}{4}(\gamma_{12})$$

Figure 2. Thermodynamics of the engulfment of a bacterium B by Phase 2 from Phase 1. Step 1, ΔG_1, approximates the free energy of adhesion. The net free energy change is $\Delta G_{net} = \gamma_{2B} - \gamma_{1B}$.

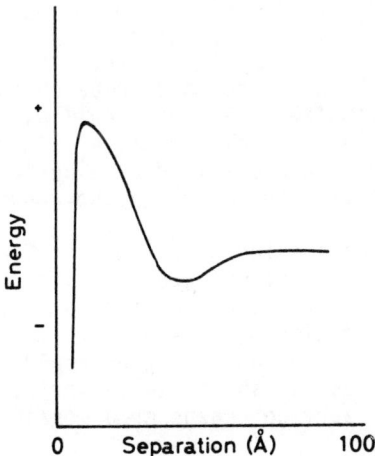

Figure 3. Net interaction energy vs. separation distance. At small separations the primary minimum surface energy effects predominate, at the secondary minimum electrostatic effects predominate.

intermolecular forces can be categorized as follows: (1) Van der
Waals forces, consisting of the attraction between a) permanent
dipoles (Keesom forces), b) dipoles induced by other permanent
dipoles (Debye forces) or c) statistical dipoles resulting from
the random motion of electrons in non-polar materials (London
forces). All of these subcategories of Van der Waals forces have
some importance in the adherence of bacteria to their glycocalyx
and of the glycocalyx to a substratum. (2) Hydrogen bonds are
intermediate in strength between covalent bonds and most Van der
Waals forces, although they are a type of dipole-dipole interaction (about 5 kcal/mole of bonds).

Van der Waals forces are sufficient to explain observed
strengths of adhesion as well as many values predicted from measurement of surface energies. Explanation, calculation and prediction of the Van der Waals forces from molecular properties is
difficult and complex. One current approach to the problem is
particularly noteworthy since it allows calculation of long-range
Van der Waals forces from data which, if not already available,
are obtained with relative ease. Parsegian (19, Parsegian and
Gingell, 20) has combined theories of molecular interaction (e.g.
DLVO theory) and more recent developments in statistical mechanics
(Lifshitz, 21) to provide a relatively comprehensive approach to
the calculation of Van der Waals forces from electromagnetic spec-
tral data covering the range from microwaves to the infrared to
the ultraviolet. These results allow calculation of the attraction between materials separated by one or more intermediate
phases and thus are applicable not only to the idealized attachment of cells to solid substrates by an adhesive, but also to the
more realistic attraction of multi-coated cells to adsorbed layers
on solid substrates (Ninham and Parsegian, 22). Parsegian and
Gingell (20) have made calculations of the attractive interactions
between hypothetical cells (hydrocarbon membrane coated by a polysaccharide layer) and actual inert substrata. Unlike calculations
based solely on measured surface energies (^{Y}c), these electrodynamic interactions can be calculated for and show differences between solids having zero contact angles. At a range of 50 Å, the
electrodynamic attractive forces calculated by this method for a
variety of surfaces are given in Table 2. These are compared to
the calculated free energy of adhesion (Eq. 8,9) for a cell with
a surface energy of 68 dynes/cm in a dilute salt solution. There
is a general correlation between the values obtained by experi-
mentation and values obtained by calculation, although the conditions are slightly different.

Several problems exist in the application of calculated Van
der Waals forces to microbial adhesions. Firstly, calculation of
the relation between net attraction and distance fails at small
distances (Fig. 3), and it is in the region of this primary minimum that calculations based on measured surface energies apply.
Secondly, this type of calculation does not specifically include
the effects of surface charges: net surface charge or a mosaic of

surface charges on the substratum and cell surface can result in either attraction or repulsion depending on charge distribution and cell geometry. In all probability, details such as these can be added to Parsegian's analysis to improve its applicability to specific biological adhesion problems.

Less detailed approaches to the calculation of surface properties from measures of polarity have also been made. The interaction parameter, ϕ, can be calculated with some accuracy from independently measured values of the dipole moment, molecular polarizability and ionization energy (Good, 6). Wu (23) has separated polar and dispersive components of interfacial and surface tensions and has demonstrated that adhesion depends on a matching of the polarities of the adherent materials.

G. Electrostatic Contribution to Adhesion. Surface free energies describe adhesion phenomena once molecular contact has been achieved between the adhering phases. For a cell to come from a distance into molecular contact with a surface requires consideration of long-range forces which influence approach. Two factors are of prime importance to the attachment of microbial cells to solid surfaces: 1) electrostatic interactions and 2) fine surface projections.

The surface of ionizable substances attracts ions of opposite charge. A monolayer of oppositely charged ions may be insufficient to neutralize the surface and further layers may build up. These ions will also attract free ions having charge of the same sign as the surface. Thus there is a diffuse cloud of ions concentrated around an ionizable surface or particle. In an electric field, this cloud will shear at some radius, leaving the particle and its cloud with a net charge. This net charge is known as the zeta potential (ζ). This and other electrokinetic phenomena are discussed by Sennett and Oliver (24).

Zeta potential is determined from electrophoretic mobility (ν) with the Helmholtz-Smoluchowski equation (Eq. 13).

$$\zeta = \frac{4\pi\eta\nu}{D} \tag{13}$$

where: ζ = zeta potential
         η = viscosity of liquid
         D = dielectric constant of liquid

From the zeta potential it is possible to obtain an approximate value for the surface charge density (Abramson et al, 25). Clearly, cells will be attracted to surfaces of opposite zeta potential. If cells have the same zeta potential as a surface, attachment is still possible provided the electrostatic barrier can be penetrated by small surface projections (van Oss et al, 18; Grinnell, 26). Zeta potentials and the adhesion of microbes have been reviewed by Daniels (27), Martin (28), and Marshall (29).

II. The Biophysics of Cell Adhesion

A. Surface Energy of Solid Substrata: Low Energy Surfaces.
A variety of studies with both low and high energy surfaces have
demonstrated that microbial attachment to solids has 2 important
aspects: the quantity of cells adhering and the strength of the
adhesion. Low energy surfaces tend to quickly accumulate large
numbers of poorly attached cells, while high energy surfaces
slowly accumulate small numbers of more firmly attached cells
(DiSalvo, 30, DiSalvo and Daniels, 31). Variation from this behaviour appears to be correlated with factors other than the surface energy of the solid.
Fletcher (32) studied the attachment of a marine pseudomonad
(NCMB-2021) to polystyrene ($\gamma c \simeq 33$ ergs/cm^2) and found that the
density of adherent cells depended on the condition of the bacterial suspension. Adhesion followed a Langmuir adsorption iso-
therm (Eq. 14).

$$\theta = k'[x]_a = \frac{k[x]}{k' + k[x]} \tag{14}$$

where: θ = fraction of surface covered by bacteria
          k' = proportionality between θ and $[x]_a$
          k = adsorption coefficient
          $[x]$ = concentration of bacteria in suspension
          $[x]_a$ = number of adsorbed bacteria per unit area
                   $(100\mu^2)$

In general, k will depend on properties of the bacterial surface
(e.g. γc) and will regulate the rate at which bacteria become
attached to the surface, while k' will depend on properties of the
solid surface and will be related to the rate at which bacteria
desorb from the surface. For polystyrene and this pseudomonad, k'
was around 2.5×10^8; k was around 5.0×10^8 but this depended
strongly on culture age. Similar results were obtained by Floodgate (33) in studies of marine *Arthrobacter* species. From his
data it can be calculated that k' for this organism is around 7 x
10^3 (for $[x]_a$ in 0.01 mm^2 units), and k is around 28×10^3. The
normal concentration of this organism in the region it inhabits is
in the order of 5×10^3 cells/ml, or slightly below its k' for
attachment. Since this attachment occurred in nutrient free
media, the importance of physical aspects of the cell-glass interaction are probably more important than are aspects of the adsorption of a film of highly concentrated nutrients.
In a further study of *Pseudomonas* NCMB 2021, Fletcher and
Loeb (34) compared attachment to a variety of low energy surfaces.
Their data, together with data from Zisman (3, 35), Metsik (36)
and Mittal (13) is plotted in Figure 4. This curve is very similar to curves relating γc or W_a to adhesive strength for plastics
(Dyckerhoff and Sell, 37) and metals (Kitazaki and Hata, 11). The

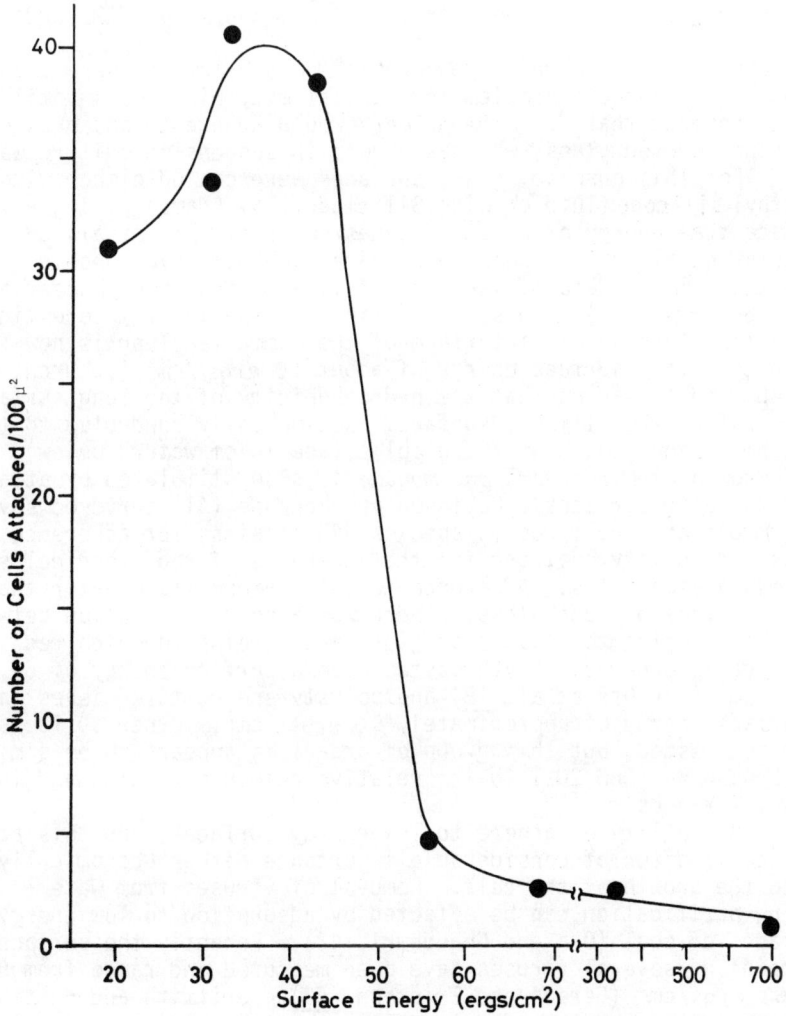

Figure 4.   Adhesion of Pseudomonas NCMB-2021 *vs. surface energy of solid substrate. See text.*

results obtained with these materials indicate that adhesive
strength increases up to approximately the point at which the
surface free energies of substrate and adhesive are equal, and
then decreases. This situation corresponds in part to the criteria for maximal adhesive strength discussed above, however in
this case the surface free energy of *Pseudomonas* sp. NCMB 2021 is
unknown.

Growth of the highly hydrophobic *Mycobacterium tuberculosis*
posed a considerable problem for medical microbiologists until it
was discovered that *M. terberculosis* would adhere to and grow on
silicone covered glass, whereas growth in suspension culture was
poor. For this purpose, glass surfaces were coated either with
dimethyl silicone fluid or with Siliclad (Clay-Adams) giving a
surface free energy of about 20 dynes/cm^2 (Higashi, et al, 38,
Tsukuma, et al, 39). *Mycobacteria* are much more hydrophobic ($\gamma_c \simeq$
16 ergs/cm^2, see data of van Oss, et al, 18) than most bacteria
($\gamma_c \simeq$ 68 ergs/cm^2). This observation is especially interesting
since the alveolar air interface of the mammalian lung is now
known to have a surface energy of about 10 ergs/cm^2 (Schurch, et
al, 40) and it may be that the hydrophobicity of the lung surface,
like that of the silicone surface, is especially conducive to the
attachment and growth of these cells (see γ_c of water, below).

Growth of *Mycoplasma pneumoniae* is also stimulated by attachment to solid surfaces. Robinson and Manchee (41) surveyed several mycoplasma cultures (9 species, 43 strains) for adherence to
glass and polystyrene, and for the adherence of red blood cells to
the mycoplasma cells. Adherence to polystyrene was greater than
that to Pyrex or soda glass. There was a poor correlation between
strains adhering to plastic or glass and strains to which red
blood cells adhered. Erythrocytes have a surface energy of 67.7
ergs/cm^2 (van Oss et al, 18) and polystyrene culture dishes have
a surface energy of approximately 56 ergs/cm^2. Other surfaces
were not tested, but this group of organisms appears to be similar
to *Pseudomonas* NCM 2021 in its relative response to high and low
energy surfaces.

Many cell types adhere to low energy surfaces, and this adherence is often of considerable importance either economically
or to the growth of the cell. Removal of viruses from water
during purification can be effected by adsorption to low energy
surfaces of coal (Oza and Chaudhuri, 42). Recently the surface
energies of several viruses have been measured and range from 68
to 63 ergs/cm^2 (Gerson and Erickson, 43). Griffith and Bultman
(44) have measured the adhesive strength of barnacles to aluminum,
PTFE-fluoroepoxy and PTFE-fluoropolyurethane surfaces. Measured
values of adhesion decreased as the surface energy decreased
(Aluminum: 6.7 kg/cm^2; PTFE-Epoxy: 4.0 kg/cm^2; and PTFE-fluoropolyurethane: 1.1 kg/cm^2). An attempt to decrease the adhesion
of marine microbes to glass surfaces by silanizing the surface to
reduce the surface energy resulted in greater numbers of microbes

adhering to the surface, but less tenaciously than with untreated
glass (DiSalvo and Daniels, 31).

Cellulose has a surface free energy of about 40 ergs/cm^2,
and values for natural woods vary between about 40 and 58 (Marian,
45, Herczeg, 46, Nguyen and Johns, 47). Latham et al (48) have
studied adhesion of *Ruminococcus flavefaciens* to cotton cellulose
and to the cell walls of rye grass (*Lolium perenne*). This bacterium selectively adheres to cell walls from the epidermis,
sclerenchyma, phloem and mesophyll, but not to bundle sheath
cells, metasylem, protoxylem, cuticle or protoplasts. *R. flavefaciens* produces an adhesive glycocalyx composed of a glycoprotein
containing rhamnose, glucose and galactose. Adhesion was a prerequisite for degradation of the plant cell wall material. Results for various types of woods and for early as opposed to late
wood indicate that there may be considerable variation in the surface free energies of the different cell types in plant tissues.

B. Adhesion to High Energy Surfaces: Glass and Clays.

1. Glass. Glass is a high energy surface ($\gamma_s \simeq 700$
ergs/cm^2) which often serves as a substrate for the adherence of
microorganisms. Certain aspects of this adhesion undoubtedly
depend on both the formation of surface films of either bound
water or organic molecules prior to colonization (see below) or
on the negative surface charge usually present, however, initially
discussion will be limited to the macroscopic observation of adhesion.

Nordin, Tsuchiya and Fredrickson (49) studied the adhesion
of *Chlorella ellipsoidea* to glass surfaces by measuring the force
required for dislodgement. *C. ellipsoidea* cells were allowed to
attach to the glass surface of a thin chamber between 2 glass
plates containing an appropriate salt solution. The relation between liquid flow rate through the chamber and the force on a
sphere in the stream is given by a modification of Brenner's (50)
equation (D>>a) (Eq. 15):

$$Fh = \frac{36\,\pi\mu\,a^2 w}{D^2 L} = 7.6 \times 10^{-6} w \qquad (15)$$

where: Fh = hydrodynamic force on the cell
        D = thickness of chamber
        L = width of chamber
        w = volumetric flow rate through chamber
        μ = fluid viscosity
        a = cell radius

This force was compared to the calculated electrostatic forces
which could be responsible for adhesion. With NaCl concentrations
ranging from 0.005 M to 2.0 M, adhesion steadily increased with
increasing salt concentration. Using FeCl resulted in more complex behaviour: at FeCl$_3$ concentrations below 5×10^{-6} M, little

adhesion was observed; between 2×10^{-5} and 1×10^{-4} M, increased adhesion was observed, up to about 8×10^{-5} dynes/cell; between 10^{-4} and 10^{-3} M, adhesion dropped to around 7×10^{-6} dynes/cell, but at 5×10^{-2} M adhesion was in excess of 2×10^{-4} dynes/cell. These results were largely explicable in terms of the electro-static attraction between the *Chlorella* cells and the charged glass surface.

Ou and Alexander (51) studied the effects of glass microbeads (600 μ - 30 μ diam) on the growth and respiration of *Bacillus megaterium*. Glass beads had a pronounced effect on the morphology of *B. megaterium*, increasing cell length from 1-5 μ to 20-600 μ. The smallest beads (30 μ) stimulated oxygen consumption by a factor of 2. Similar results have been observed for the effects of clay particles on microbial respiration (see below). Increased oxygen utilization corresponded to increased glucose utilization. As with *Chlorella*, the negatively charged surface was an important aspect of the observed effects: modification in Mg^{++} requirements were observed in the presence of glass beads and the elongation of cells is often associated with Mg^{++} deficiency. Taylor and Parkinson (52) studied *Penicillium decumbens* in glass microbead (1-350 μ) culture. Respiration (CO_2 evolution) was stimulated by beads in the 250-350 μ size class relative to the 100-150 μ or 0-30 μ groups. This organism dwells in sand dunes and it was found that the glass microbeads simulated its preferred habitat.

Marine microbes are frequently found attached to surfaces, or will attach to a surface given the opportunity. Early studies of marine fouling of glass surfaces were conducted by Zobell (53) and Kriss (54). One of their most important observations was that the adsorption of organic materials by solid surfaces concentrated microbial nutrients and that this led to colonization. The work of Floodgate (33) and others indicates that while adsorbed layers may stimulate growth, purely physical factors can also dominate the interaction.

2. Clays. The finest inorganic particulates of most soils are particles of clay minerals. Clays are commonly layered, plate-like structures composed of aluminosilicates and possessing highly variable surface and hydration properties. The most common types of clay are montmorillonite, kaolinite and illite. All are mixed, layered structures containing sheets of silica and alumina: kaolinite has a 1:1 ratio of silicon to aluminum sheets, montmo-rillonite has a 2:1 ratio, with a sandwich-like structure, and illite also has a 2:1 structure, but contains more aluminum and potassium cations than does montmorillonite. In an aqueous envi-ronment, kaolinite is relatively stable and has a low cation ex-change capability (10-100 μeq/g), while montmorillonite swells markedly and has a high cation exchange capacity (800-1200 μeq/g). Illite clays have somewhat intermediate properties. As a soil constituent, and as a solid substrate for microbial attachment and growth, clays can offer an enormous surface area (2-3 m^2/g) and

the cation exchange capability can provide a rich source not only
of mineral nutrients but also of organic nutrients to attached
microbial cells.

Microbes spontaneously attach to clay particles in the environment in large numbers. Early work by Minenkow (55) demonstrated that microbes are attached or adsorbed by some means to
clay particles in the soil. For both *Bacillus cereus* and *Serratia
marcescens*, the number of adsorbed cells was a function of the
clay content of a wide variety of soil compositions. Scanning
electron microscopy (SEM) has demonstrated that large numbers of
microbes are firmly attached to suspended inorganic particulates
found in both fresh and salt waters. The modes of attachment
include adhesion by capsular secretions or microexudates, fibril-
lar or polymeric holdfasts and indiscernable attachment mechanisms which presumably do not depend on an intermediary adhesive.
Paerl (56) studied oceanic surface particulates (off southern
California) and found that clays and diatomaceous skeletons were
coated with slimes and a multitude of attached bacterial species.
In fresh waters (Lake Tahoe) suspended mineral particles supported
colonies of bacteria attached by networks of fine fibrils. Controlled studies of bacterial colonization were performed with
fresh silt particles retained in dialysis bags and incubated in
Lake Tahoe for 2 weeks. The particulates became heavily colonized
by microbes, with a 2:1 ratio of attached to free cells. An important component of the attachment adhesive was acid polysaccharide, with lesser quantities of protein.

Microbial attachment to rocks and sand particles forms the
basis for the largest and most extensive utilization of immobilized microbial cells in the world today: trickling filters.
Mack et al (57) studied the development and ecological succession
of microbial films in a newly commissioned trickling filter system. A biofilm formed over the rock surfaces and consisted of a
micro-jungle of bacterial colonies, fungi, algae and protozoa.
The surface of the biofilm consists of a highly convoluted, tufted
mat of microbial biomass. Diatoms were a common constituent of
this microbial community, and their silaceous skeletons provide
new surfaces for colonization. Most of the bacterial colonies
consisted of a mixture of species and were of an open cylindrical
shape, affording maximum aeration and nutrient exposure.

The soil microbiology literature (e.g. see Hattori and
Hattori, 58) makes it clear that microbes readily adhere to clay
surfaces and that clays will often provide stimulus for both
growth and metabolic activity. The use of clay surfaces (ceramics) for processes involving immobilized whole cells has recently
been investigated by Marcipar et al (59). *Saccharomyces cerevisiae, Candida tropicalis, Trichosporon* sp. and *Rhodortorula* sp.
were studied for their degree of attachment to a ceramic substrate
(Biodamine) at pH 4 and pH 6. At pH 4, 40-75% of cells exposed
to the surface adhered strongly enough to resist removal by a
flowing stream of medium (linear flow rate 2 cm/min). The

strength of attachment was determined using the method of Nordin
et al (49) (see above). For *S. cerevisiae* and *C. tropicalis*, pH
had only a small effect on attachment as measured by linear flow
rate: the average maximum sustainable flow for both cultures was
4.1 cm/min. Respiration rate was determined for free cells and
for cells attached to the ceramic substrate, and the ratio at pH
4 for fixed:suspended cells was 6.7. One probable explanation of
the improved metabolic activity of immobilized cells is the crea-
tion of an ideal microenvironment around each cell through local
adsorption phenomena and the prevention of glycocalyx loss which
is common in suspension culture (Costerton et al, 60).

        C. Surface Energies of Living Cells. There are few measure-
ments of the surface energies of living cells. The largest series
of measurements is that of van Oss et al (18) which were obtained
from the contact angles of normal saline on dried microbial
films. Similar measurements have been made by Gerson and Zajic
(66) for *Corynebacterium lepus* (see below). This technique has
the advantage of simplicity but has the difficulty that the cell
surface may change irreversibly during drying. Harvey and
Marsland (61) measured the surface tension of *Amoeba dubia* by
observing the stretching of the cell surface by cytoplasmic
inclusions in a centrifugal field. The interfacial tension at
the cell surface in growth medium was 1-5 dynes/cm. Interfacial
tension between oil and protoplasm was also found to be about
1 dyne/cm (Harvey and Shapiro, 62). The surface tension of alveolar cells in the rat lung was measured by Schurch et al (40) and
was found to be in the range of 9-11 dynes/cm. Table 3 summarizes
some data on cell surface energies.

        D. Utilization of Insoluble Substrates. The adhesion of
microbes to both solid and liquid hydrocarbons is a well-known
phenomenon and it is well established that attachment to hydrocarbon droplets or fragments is essential to oxidation and metabolism. It is remarkable that the surface physics of this situation has been ignored, considering the importance of hydrocarbon
fermentations in large-scale single-cell protein production (63)
and their interest to the field of biochemical engineering. Recently measurements have been made of the surface energy of *Corynebacterium lepus*, which is an aggressive utilizer of both liquid
and solid alkanes between dodecane and eicosane (Gerson and Zajic,
64, 65, 66). This organism, as well as many other hydrocarbon
utilizing microbes, produces surface-active materials which result
in both emulsification and wetting of the hydrocarbon. Interfacial (growth medium-hydrocarbon) and surface tensions have been
studied through the growth period of *C. lepus*, and at approximately 40 hours, interfacial tension reaches a minimum. At around
this time, cells of *C. lepus* move from being mostly in the aqueous
phase to being mostly in the hydrocarbon phase. Analysis of this
engulfment of the microbial cells by the hydrocarbon droplets

Table III

Surface Tensions of Cells and Microorganisms

Cell or Microbe	γ_c (ergs/cm^2)	Reference
Thiobacillus thiooxidans	≥72	66
rat neutrophil	68.3	18
guinea pig neutrophil	68.1	18
hela cells	67.8	18
mouse neutrophil	67.7	18
human erythrocytes	67.7	18
human neutrophil	67.4	18
Staphylococcus aureus-Smith	67.3	18
Streptococcus pneumoniae-1	67.1	18
chicken neutrophil	67.1	18
human platelets	67.1	18
E. coli-0111	67.0	18
rabbit neutrophil	66.9	18
Haemophilus influenzae-B	66.9	18
human fibroblasts	66.8	18
monkey neutrophil	66.8	18
human neutrophil	66.8	18
human polymorphonuclear leucocytes	66.8	18
Staphylococcus aureus	66.5	18
Salmonella arizonae	66.4	18
Salmonella typhimurium	66.0	18
E. coli-07	64.9	18
Listeria monocytogenes (rough)	63.4	18
Nesseria gonorrhoeae	63.4	18
Brucella abortus	63.2	18
Pseudotsuga menziesii (xylem)	59-53	47
Western hemlock (xylem)	56.5	47
Cornyebacterium lepus	48.7	66
English oak (xylem)	40.8	47
human skin	38.2	47
Mycobacterium butyricum	15.6	18

using the methods described by Neumann et al ($\underline{7}$, $\underline{8}$) is as follows: the net free energy change for the engulfment of the bacterium by the hydrocarbon phase is -10.88 ergs/cm^2 (both ΔG_1 and ΔG_2 are negative). For the engulfment of hydrocarbon by the bacterium at this point in the fermentation, ΔG net is -1.5 ergs/cm^2 (ΔG_1 is negative and ΔG_2 is positive, implying that hydrocarbon will attach to or wet the bacterial surface but will not be engulfed in the sense of step 2, Fig. 3). These calculations are based on conditions of minimal interfacial tension. Initially, oil-water interfacial tension is higher (25 dynes/cm), and under these conditions ΔG net is -25 ergs/cm^2 (both stages are spontaneous). Continual decrease in the oil-water interfacial tension during the fermentation, resulting in an increase in ΔG net, could be an important aspect of the linear growth often observed in hydrocarbon fermentations. However, while ΔG net is increasing, a decrease in interfacial tension between the medium and hydrocarbon results in increased emulsification and increased hydrocarbon surface area available to the microbes, which should stimulate net uptake by the population.

Colonization of solid surfaces by *C. lepus* has also been investigated in terms of solid surface free energies. Figures 5 and 6 are scanning electron micrographs of the colonization of aluminum and polyethylene respectively by cells of *C. lepus*. The highly hydrophobic microbe shows a strong affinity for hydrophobic surfaces.

Thiobacillus thiooxidans colonizes elemental sulfur and oxidizes it to sulfuric acid. In this process, *T. thiooxidans* produces wetting agents which have been variously identified as phosphatidyl inositol, phosphatidyl glycerol, phosphatidyl ethanolamine (Schaeffer and Umbreit, $\underline{67}$; Jones and Benson, $\underline{68}$) or diphosphatidyl glycerol (Agate and Vishniac, $\underline{69}$). Sulfur is a low energy surface with a surface energy of approximately 31 ergs/cm^2 ($\underline{70}$). During growth, cells adhere to the sulfur and etch the surface by an unknown mechanism (Schaeffer et al, $\underline{71}$). The surface energy of *T. thiooxidans* has been measured (Gerson and Zajic, $\underline{66}$) and found to be greater than or equal to 72 ergs/cm^2, and this value leads to a ΔG net of -.35 ergs/cm^2 for the engulfment of sulfur by the bacterium (both steps spontaneous). While engulfment may not be the mechanism by which *T. thiooxidans* gains metabolic access to particles of elemental sulfur, this calculation indicates that attachment to sulfur and its movement into the cell can be thermodynamically spontaneous. Colloidal or molecular sulfur particles may diffuse into the cell or be dissolved by hydrophobic cell constituents. The free energy of adhesion of *T. thiooxidans* to sulfur is -73 ergs/cm^2 or less. *T. thiooxidans* also attaches to oil shale (Yen, $\underline{72}$). The surface energy of oil shale has also been measured (Gerson and Zajic, $\underline{66}$), and the free energy of attachment to and engulfment of oil shale by *T. thiooxidans* is negative, as is the free energy of adhesion. Although the appropriate surface energies have not yet been meas-

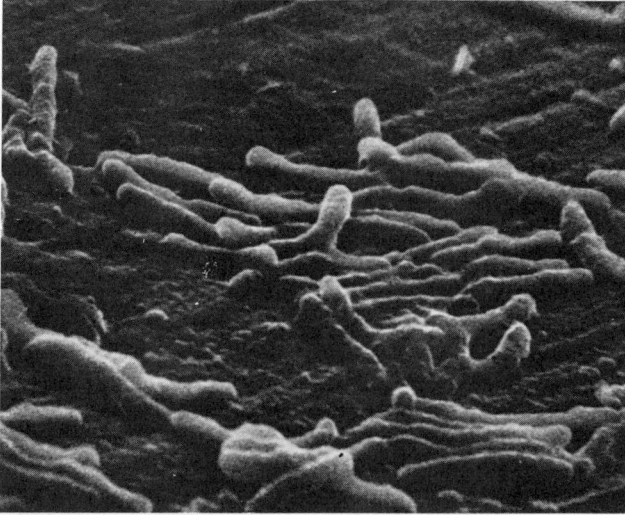

Figure 5. Corynebacterium lepus *attached to aluminum, bar indicates 1.0 μ.*

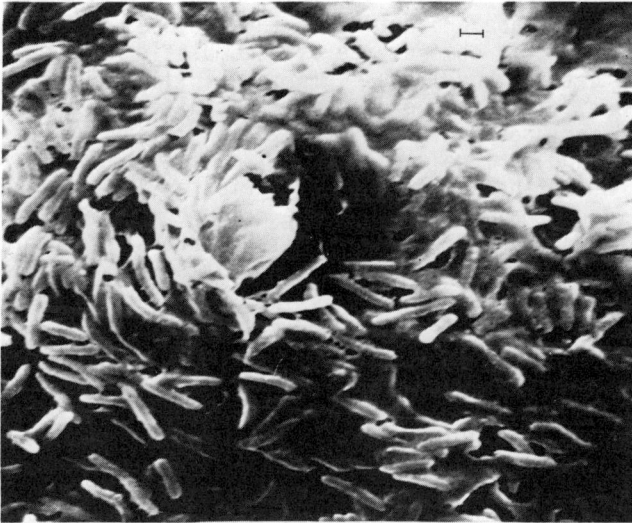

Figure 6. Corynebacterium lepus *attached to polyethylene, bar indicates 1.0 μ.*

ured, it is observed that *T. thiooxidans* and *T. ferroxidans* adhere
to pyritic surfaces, possibly through a lipopolysaccharide (Duncan et al, 73, Hirt and Vestal, 74). Use of *T. thiooxidans* in
industrial heap leaching of uranium (see Murr et al, 75 and
Schwartz, 76) is in effect a very large immobilized cell bioreactor for the production of sulfuric acid from pyritic materials
or sulfur.

Similar analyses should be performed for other microbes which
adhere to and utilize solid or immiscible substrates (e.g. cellulose, lignin, chitin), since calculation of ΔG net for a proposed
mechanism of utilization would identify likely processes, and
dissection of the mechanism into stages would identify the most
probable routes used by the microbe.

E.   Influence of Interfacial Films on Microbial Attachment.
Air-water and solid-water interfaces accumulate surface-active or
hydrophobic organics present in the aqueous phase. Both the increased concentration of organic foodstuffs and the potentially
increased adhesiveness of the interfacial film tend to increase
the relative concentration of microbes at interfaces. In terms
of cell immobilization, it may be possible to increase the
strength of adhesion and the durability of the immobilized whole-
cell system by pre-coating the solid substrate with an appropriate
interfacial film.

Ecological interfacial films range from molecular monolayers
(~20Å) to a millimeter or more; microbial cells are typically
1-10 μ in diameter and are thus ideally proportioned to take simultaneous advantage of both the bulk phases and interfacial
regions. The surface properties of the microbes themselves will,
of course, result in some degree of accumulation at the interface,
where under natural conditions interfacial films will aid growth
or attachment. The biota of the interfacial ecosystem is termed
neuston and includes microbes, protozoa and algae.

Filming and surface settlement of marine larvae has been
studied by Crisp and Ryland (77) and Crisp and Meadows (78), and
they have found that a water soluble component of arthropod
cuticle, arthropodin, greatly facilitates adhesion to solids.
Loeb and Neihof (79) studied the accumulation of organics on
platinum in the estuarial waters of Chesapeake Bay and found not
only that there was a steady accumulation of material but also that
this organic material greatly reduced the solid surface energy as
well as the zeta potential of the platinum interface (+0.55 to
-0.79 μM-V/cm-sec).

The neuston of marine interfaces has been studied extensively. At the air-water interface, there is an accumulation of
higher molecular weight, hydrophobic molecules including chloro-
phylls, xanthophylls and carotenoids (Parker and Barsom, 80).
The relative concentrations of organisms at a depth of 10 cm compared to the air-water interface were measured by Harvey (81) and
found to be as follows: colorless flagellates, 0%; dinoflagel-

lates, 12.5%; cillates, 112%; and diatoms, 247%. These results indicate that certain microbial populations are concentrated at the interface while others primarily reside in the bulk water. DiSalvo (30) studied the coating of glass surfaces by bacterial neuston present at the air-ocean interface. Bacterial numbers picked up at the glass-air-water interface compared to those collected at the glass-water interface were in the order of 1000:1, whereas the bulk concentration of bacteria in the top 0.3 mm compared to several centimeters below the surface were in the order of 10:1. These results indicate that there is considerably greater adhesion to glass at the 3-phase interface than at the 2-phase interface. This may be due to adhesive microexudates produced by the bacteria on contact with a solid surface (Marshall et al, 82), although a very considerable amount of attachment occurs in a very short time (1-2 min) and thus it is unlikely to depend on physiological response which is initiated by contact with a solid. Concentration of surface films at the interface confers ecological advantage and certain microbial species may produce a sticky coat while suspended in the aqueous phase which allows rapid and firm attachment to virtually any solid surface with which they collide. Repeated collisions would be expected to result in aggregates of microbes and inert particles (DiSalvo, 1973). Unlike the results obtained for glass, bacterial adhesion to agar surfaces and to liquid films on agar surfaces was found to depend simply on the concentration of bacteria in the bulk medium. While bubbles are well known to concentrate bacteria at the air-water interface, study of microbial adhesion in 3-phase systems (gas-liquid-solid) is lacking at the present time.

Plant and other lectins have the ability to adhere strongly to microbial cell surfaces and lectins are responsible for the adhesion of *Rhizobium trifolii* to root hair cells of clover. Discovery of lectins which will bind other microbes may lead to the use of lectins as bridging agents to bind cells to inert support materials. Fletcher (83) studied the effects of proteins on the adhesion of a marine *Pseudomonas* sp. to polystyrene Petri dishes. Bovine serum albumin, gelatin, fibrinogen and pepsin ($pK_I \leq 5.8$) all inhibit attachment at pH 7.6 when present either prior to or concurrently with the microbial cells. Bovine serum albumin decreased the adhesion of previously attached cells. Basic proteins protamine ($pK_I = 12.4$) and histone ($pK_I = 10.8$) did not inhibit attachment. The effect of proteins on attachment were independent of surface charge density on the substratum and thus the decreased attachment in the presence of proteins may be due to non-electrostatic interactions. The behaviour of bovine serum albumin, which has a large number of non-polar side chains, indicates that hydrophobic interactions may be important (Tanford, 84; Goldsack and Chalifux, 85). The effects of hydrophyllic colloids on bacterial flocculation have been studied by Hodge and Metcalfe (86) and the subject has been reviewed by Harris and Mitchell (87).

Zisman (35) has recently studied the effects of adsorbed water on the surface energy of metals, quartz and other high energy surfaces. As the degree of surface hydration increased, the surface behaved more like the surface of water, with a surface energy of 22 ergs/cm^2. For instance, the surface energy of a hydrated glass surface is about 30 ergs/cm^2, while at very low relative humidities (RH), glass behaves like a higher energy surface, γc = 45 ergs/cm^2. Similar results are obtained with metal surfaces: at 0.6% RH, γc is about 45 ergs/cm^2 and at 95% RH, γc is about 37 ergs/cm^2 for a series of 13 metals. The uniformity of the values obtained for γc on a wide variety of high energy solids indicates that in practice, the value of γc for high energy solids will be relatively low because of the adsorbed water layer. This low value of 30-45 ergs/cm^2 is probably the value applicable to problems involving microbial attachment unless the microbe is able somehow to dehydrate the surface at the point of attachment.

F. Adhesion and Cells from Higher Organisms. The qualitative aspects of the adhesiveness of mammalian cells has been reviewed recently by Grinnell (26). From the qualitative viewpoint, mammalian cells adhere to solid surfaces by an extracellular layer of protenaceous microexudate, and attachment to protein-free surfaces is non-specific, in the sense of enzyme specificity, but still depends to some extent on the chemical composition of the substrate. Cells contact the substratum through cytoplasmic microextensions or filopodia which are able to pierce the electrostatic barrier because of the small actual force exerted on them (both cells and many substrata carry a net negative charge; see also van Oss et al (18)). Binding to protenaceous surfaces (e.g. other cells) can be highly specific. Attachment is often followed by spreading resulting from microtubule and cytoskeletal restructuring, which often conforms to any structures (e.g. grooves) present on the substratum (Ivanova and Margolis, 88). Negative surface charge densities in the order of 5 charges/100 Å^2 are required for cell growth and division. Attachment of mammalian cells to surfaces is often enhanced by divalent cations which act as a bridging adhesive (Baier et al, 89).

Quantitative aspects of the requirements of mammalian cells for surfaces have been studied by Maroudas (90, 91, 92), Taylor (93), Beukers et al (94) and Weiss and Blumenson (95). The results of these investigations are that in the absence of high protein concentrations in the growth medium, mammalian cells are sensitive to the surface energy of the substratum, but lose this sensitivity as protein (serum) concentration increases. On the average, an intermediate surface energy (e.g. 55 ergs/cm^2) and a relatively high negative surface charge density are optimal. These requirements are best explained at present by the theory that spreading and tensioning are required for cell division in mammalian cells (Curtis and Seehar, 96, Maroudas, 91).

Attachment of cells to cells is important both in pathogenesis and morphological development. Attachment of microbes to tissue surfaces has been reviewed by Costerton et al (60), Jones, Abrams and Freter (97), Tay and von Fraunhofer (98), Freter and Jones (99) and Vasilieu and Gelfand (100). Cell-cell adhesiveness in morphological development, the differential adhesiveness theory, has been studied and reviewed by Nardi and Kafalos (101, 102). Since it is known that attachment is required for mammalian cells, it is interesting that attachment characteristics have not yet been thoroughly studied for the notoriously unsuccessful culture of monocotyledenous protoplasts.

III. Conclusions

These considerations lead to the conclusion that a rational approach to problems of the adhesion of cells to solid surfaces can be developed from knowledge of the surface properties of both the substrate and the cell. Solid surface energies can be obtained by measurements of contact angles and use of Neumann's equation (Eq. 6), thus allowing calculations of free energy charges associated with adhesion. Zeta potentials and resultant electrostatic contributions to adhesion can also be obtained experimentally. This type of approach should provide insight into microbial adhesion problems in the marine and aquatic environments, disease and infection and in the industrial immobilization of whole cells.

Literature Cited

1 Jack, T.R. and Zajic, J.E. Adv. Biochem. Engr. (1977), 5, 125-145.
2 Zisman, W.A. Ind. Eng. Chem. (1963), 55, 10,18.
3 Zisman, W.A. Adv. Chem. Ser. (1964), 43, 1.
4 Young, T. Phil. Trans. Roy. Soc. (1805), 95, 65-87.
5 Good, R.J. and Girifalco, L.A. J. Phys. Chem. (1960), 64, 561.
6 Good, R.J. in: Lee, L.H. Adhesion Science and Technology (Polymer Science and Technology, 9A, 107-127), Plenum, 1975.
7 Neumann, A.W., Gillman, C.F. and van Oss, C.J. J. Electroanal. Chem. (1974), 49, 393.
8 Neumann, A.W., Good, R.J., Hope, C.J. and Sejpal, M. J. Coll. Int. Sci. (1974), 49, 291.
9 Good, R.J. Adv. Chem. Ser. (1964), 43, 74.
10 Good, R.J. and Elbing, E. Ind. Eng. Chem. (1970), 62(3), 54.
11 Kitazaki, Y. and Hata, T. J. Adhesion. (1972), 4, 123.
12 Rhee, S.K. Mat. Sci. Engr. (1973), 11, 311.
13 Mittal, K.L. in: Lee, L.H. Adhesion Science and Technology (Polymer Science and Technology, 9A, 129-168), Plenum, 1975.
14 Gray, V.R. in: Proc. 4th Intl. Cong. Surface Activity, Gordon and Breach, New York, 1967.
15 Albertsson, P.A. Endeavour (1977), 1, 69-74.
16 Larsson, D.O., Ohlson, S. and Mosbach, K. Nature (1976), 263, 796-797.
17 Neumann, A.W., van Oss, C.J. and Szekely, J. Kolloid Z. u. Z. Polymere. (1973), 251, 415-423.
18 van Oss, C.J., Gillman, G.F. and Neumann, A.W.,"Phagocytic Engulfment and Cell Adhesiveness", Dekker, N.Y., 1975.
19 Parsegian, V.A. "Long Range Van der Waals Forces", in: van Olphen, H. and Mysels, F.J. (eds) "Physical Chemistry", Theorex, La Jolla, pp. 27-72, 1975.
20 Parsegian, V.A. and Gingell, D. J. Adhesion. (1972), 4, 283-306.
21 Lifshitz, E.M. Sov. Phys. JETP. (1956), 2, 73.
22 Ninham, B.W. and Parsegian, V.A. Biophys. J. (1970), 10, 646-663.
23 Wu, S. in: Lee, L.H. "Recent Advances in Adhesion", Gordon and Breach, London, 1973.
24 Sennett, P. and Oliver, J.P. Ind. Engr. Chem. (1965), 57(8), 33-50.
25 Abramson, H.A., Moyer, L.S. and Goren, M.H. "Electrophoresis of Proteins and the Chemistry of Cell Surfaces", Reinhold, N.Y., 1942.
26 Grinnell, F. Int. Rev. Cytol. (1978), 53, 65-144.
27 Daniels, S.L. Dev. Ind. Microbiol. (1971), 13, 211-253.

28   Martin, P.   "Electrodeposition of Bacterial Aerosols". M.Sc.
     thesis, University of Western Ontario, 1978.
29   Marshall, K.C.   "Interfaces in Microbial Ecology", Harvard
     Univ. Press, Cambridge, 1976.
30   DiSalvo, L.H. Limnol. Oceanogr. (1973), 18, 165-168.
31   DiSalvo, L.H. and Daniels, G.W. Microbial Ecol. (1975), 2,
     234-240.
32   Fletcher, M. J. Gen. Microbiol. (1976), 94, 400-404.
33   Floodgate, G.D. Veroff. Inst. Meeresforsch. Bremerh. (1966),
     2, 265-270.
34   Fletcher, M. and Loeb, G.I. in: Kerker, M. (ed). Colloid and
     Interface Science (1976), 3, 459-463.
35   Zisman, W.A. in: Lee, L.H. Adhesion Science and Technology
     (Polymer Science and Technology, 9A, 55-92), Plenum,
     1975.
36   Metsik, M.S. in: Lee, L.H. (ed) "Recent Advances in
     Adhesion", Gordon and Breach, London, 1973.
37   Dyckerhoff, G.A. and Sell, P.J. Makromol. Chem. (1972), 21
     (312), 169.
38   Higashi, K., Tsukuma, S. and Naito, M. Amer. Rev. Resc. Dis.
     (1962), 85, 392-397.
39   Tsukuma, S., Imoto, G. and Naito, M. Amer. Rev. Resp. Dis.
     (1965), 91, 758-761.
40   Schurch, S., Goerke, J. and Clements, J.A. (1976), P.N.A.S.
     (USA) 73, 4698-4702.
41   Robinson, D.T. and Manchee, R.J. J. Bacteriol. (1967), 94,
     1781-1782.
42   Oza, P. and M. Chaudhori. J. Gen. Appl. Microbiol. (1977),
     23, 1-6.
43   Gerson, D.F. and Erickson, J. (1978), unpublished results.
44   Griffith
45   Marian, J.R. in: "Symposium on Properties of Surfaces".
     Astm. Mat. Sci. Ser. 4, Spec. Tech. Pub. 340 (1962),
     122-149.
46   Herczeg, A. Forest Prod. J. (1965), 15 (11), 499-505.
47   Nguyen, T. and Johns, W.E. Wood. Sci. Technol. (1978), 12,
     63-74.
48   Latham, M.J., Brooker, B.E., Pettipher, G.L. and Harris, P.J.
     Appl. Env. Microbiol. (1978), 35, 156-165.
49   Nordin, J.S., Tsuchiya, H.M. and Fredrickson, A.G. Biotech.
     Bioengr. (1967), 9, 545-558.
50   Brenner, H. Chem. Eng. Sci. (1964), 19, 703.
51   Ou, L.T. and Alexander, M. Arch. Microbiol. (1974), 101, 35-
     44.
52   Taylor, H.I. and Parkinson, D. Can. J. Microbiol. (1971),
     17, 967-973.
53   Zobell, C.E., "Marine Microbiology", Chronica Botanica Co.,
     Watham, 1946.
54   Kriss, A.E. "Marine Microbiology (Deep Sea)". English
     translation: Shenan, J.M. and Kabuta, Z. Edinburgh

and London, 1963.
55 Minenkow, A.R. Zentralb. Bakteriol. Abt. II (1929), 78, 109-
     112.
56 Paerl, H.W. Microbial Ecol. (1975), 2, 73-83.
57 Mack, W.N., Mack, J.P. and Ackerson, A.O. Microbiol. Ecol.
     (1975), 3, 215-226.
58 Hattori, T. and Hattori, R. CRC Crit. Rev. Microbiol. (1976),
     4, 423-461.
59 Marcipar, A., Cochet, N., Brackenbridge, L. and Lebeault, J.
     M. Proc. 1st. Eur. Cong. Biotechnol. (1978), 1, 178-181.
60 Costerton, J.W., Geesey, G.G. and Cheng, K.J. Sci. Am.
     (1978), 238 (1), 86-95.
61 Harvey, E.N. and Marsland, D.A. J. Cell. Comp. Physiol.
     (1932), 2, 75-97.
62 Harvey, E.N. and Shapiro, H. J. Cell. Comp. Physiol. (1934),
     5, 255-267.
63 Tannenbaum, S.R. and Wang, D.I.C. (eds), "Single Cell Pro-
     tein", MIT Press, Cambridge, 1975.
64 Gerson, D.F. and Zajic, J.E. Dev. Ind. Microbiol. (1978),
     19, 577-599.
65 Gerson, D.F. and Zajic, J.E. in: Redford, D.A. and Wine-
     stock, A.G., The Oil Sands of Canada-Venezuela, CIMM
     Special volume 17, 705-710 (1978).
66 Gerson, D.F. and Zajic, J.E. (1978), unpublished results.
67 Schaeffer, W.I. and Umbreit, W.W. J. Bact. (1963), 85, 492-
     493.
68 Jones, G.E. and Benson, A.A. J. Bact. (1965), 89, 260-261.
69 Agate, A.D. and Vishniac, W. Arch. Microbiol. (1973), 89,
     247-255.
70 Olsen, D.A., Moravec, R.W. and Osteraas, A.J. J. Phys. Chem.
     (1967), 71, 4464-4466.
71 Schaeffer, W.I., Holbert, P.E. and Umbreit, W.W. J. Bact.
     (1963), 85, 137-140.
72 Yen, T.F. "Science and Technology of Oil Shale", Ann Arbor
     Science, Ann Arbor, 1976.
73 Duncan, D.W., Trussell, D.C. and Walden, C.C. Appl. Micro-
     biol. (1964), 12, 122-126.
74 Hirt, W.E. and Vestal, J.R. J. Bact. (1975), 123, 642-650.
75 Murr, L.E., Torma, A.E. and Brierley, J.A. (eds), "Metal-
     lurgical Applications of Bacterial Leaching and Related
     Microbiological Phenomena", Academic Press, New York,
     1978.
76 Schwartz, W. (ed), "Bacterial Leaching", Verlag Chemie, New
     York, 1977.
77 Crisp, D.J. and Ryland, J.S. Nature (1960), 185, 119.
78 Crisp, D.J. and Meadows, P.S. Proc. Roy. Soc. B. (1963),
     158, 364-387.
79 Loeb, G.I. and Neihof, R.A. J. Marine Rsch. (1977), 35, 283-
     291.
80 Parker, B. and Barsom, G. Bioscience (1970), 20 (2), 87-93.

81 Harvey, G.W. Limnol. Oceanogr. (1966), 11, 608-613.
82 Marshall, K.C., Stout, R. and Mitchell, R. J. Gen. Micro-
       biol. (1971), 68, 337-348.
83 Fletcher, M. Can. J. Microbiol. (1977), 23, 1-6.
84 Tanford, C. Science (1978), 200, 1012-1018
85 Goldsack, D.E. and Chalifoux, R.C. J. Theor. Biol. (1973),
       39, 645-651.
86 Hodge, H.M. and Metcalf, S.N. J. Bact. (1958), 75, 258-264.
87 Harris, A.H. and Mitchell, R. Ann. Rev. Microbiol. (1973),
       27, 27-50.
88 Ivanova, O.Y. and Margolis, L.B. Nature (1973), 242, 200.
89 Baier, R.E., Shafrin, E.G. and Zisman, W.A. Science (1968),
       162, 1360-1368.
90 Maroudas, N.G. Exp. Cell. Res. (1973), 81, 403.
91 Maroudas, N.G. Nature (1973), 244, 353-354.
92 Maroudas, N.G. J. Theor. Biol. (1975), 49, 417-424.
93 Taylor, A.C. Exp. Cell. Res. (1961), 8, 154-173.
94 Beukers, H., Deirkauf, F.A., Blom, C.P., Dierkauf, M. and
       Riemersma, J.C. J. Cell. Physiol. (1978), 97, 29-36.
95 Weiss, L. and Blumenson, L.E. J. Cell. Physiol. (1968),
       70, 23-32.
96 Curtis, A.S.G. and Seehar, G.M. Nature (1978), 274, 52-53.
97 Jones, G.W., Abrams, G.D. and Freter, R. Inf. and Immunol.
       (1976), 14, 232-239.
98 Tay, W.M. and von Fraunhofer, J.A. Surf. Technol. (1978),
       7, 157-163.
99 Freter, R. and Jones, G.W. Inf. and Immunol. (1976), 14,
       246-256.
100 Vasilieu, J.M. and Gelfand, I.M. Nature (1978), 274, 710-
       711.
101 Nardi, J.T. and Kafalos, F.C. J. Emb. Exp. Morph. (1976),
       36, 469-487.
102 Nardi, J.T. and Kafalos, F.C. J. Emb. Exp. Morph. (1976),
       36, 489-512.

RECEIVED March 1, 1979.

Production of Antibiotics and Enzymes by Immobilized Whole Cells

SHUICHI SUZUKI and ISAO KARUBE

Research Laboratory of Resources Utilization, Tokyo Institute of Technology, 4259 Nagatsuta-cho, Midori-ku, Yokohama 227, Japan

Recently many microorganisms have been immobilized by various methods (1,2,3) and immobilized whole cells have been used for industrial production (4,5). A possible advantage of whole microorganisms relative to purified enzymes is the ability to catalyze a series of linked reactions, some of which require cofactors. Continuous production of hydrogen and microbial electrodes using immbobilized living whole cells have been reported by the authors (6-11). It was found that whole cells in gel matrix and collagen membrane lived for a month under certain conditions. Immobilized living whole cells have a potential application to a new type of fermentation. Production of biologically active substances can be performed with a column system using immobilized whole cells. In this case, a fermentation process can be controlled easily. Furthermore, processes for separation of the end products can be connected directly with a column system. In this paper, penicillin G production by immobilized whole cells of Penicillium chrysogenum, bacitracin production by immobilized whole cells of Bacillus sp. and, ⍺-amylase production by immobilized whole cells of Bacillus subtilis are described.

Penicillin G Production by Immobilized Whole Cells

A few reports have appeared in the literature on the use of living immobilized whole cells for producing traditional fermentation products (3,12). In this paper, our work on the immobilization of Penicillium chrysogenum ATCC 12690 in polyacrylamide gel, collagen membrane, and calcium alginate and the production of penicillin G from glucose by the immobilized mycelium are described.

Screening of Carriers for Immobilization. Washed mycelium produced penicillin G similar to the results described by Halliday and Arnstein (13) when it was resuspended in an incubation medium. Under similar conditions, the average rate of penicillin production was comparable to that in the fermentation described above [usually between 10 and 13 units/ml h]. However, the activity of penicillin G production decreased rapidly with repeated use. As lysis of the washed mycelium was observed during these experiments, the multi-enzyme system in the mycelium for penicillin synthesis might have leaked out of the cells.

Penicillium chrysogenum was immobilized in collagen membrane, calcium alginate gel, and polyacrylamide gel. Penicillin production by different immobilized mycelia are compared in Table 1. The mycelium immobilized in calcium alginate gel

Table I   Penicillin production by
          immobilized mycelium

	Carrier	Activity (%)
Washed mycelium	– – –	100
Immobilized mycelium	Collagen	(trace)
	Calcium alginate	45
	Polyacrylamide	15

showed the highest activity. However, the gel prepared from calcium alginate became too fragile in the presence of phosphate ion. Therefore, it was difficult to use it for penicillin production repeatedly. The activity of the mycelium in collagen membrane was substantially lower than that in other carriers. This might be due to the inactivation of the mycelium with the tanning reagents, viz., glutaraldehyde. The mycelium immobilized in polyacrylamide gel showed fairly good activity. However, the rate of penicillin production of the immobilized mycelium was lower than that of washed mycelium. This might arise from the inactivation of the mycelium by polymerization reagents, such as monomers, APS, and TEMED. The reagents, such as acrylamide and TEMED, severely and adversely affected the penicillin-producing activity by the washed mycelium. These results suggested that the multi-enzyme system required for penicillin synthesis was inactivated by the reagents individually. Therefore, the amounts of acrylamide and TEMED were decreased to as low as possible. Five percent of acrylamide and 0.05 % TEMED were used for polymerization of the gel.

The total concentration and the relative ratio of acrylamide

and BIS determine both the characteristic pore size of the gel
in which whole cells are entrapped as well as the physical
properties of the finished gel. The effect of BIS content in
monomers on penicillin production by the immobilized mycelium
was examined. The activity increased with increasing BIS
content. Since the gel was fragile when the BIS content in the
monomer was higher than 20 %, a BIS content of 15 % was employed
for further work.

Effect of Substrates on Penicillin Production. It is known
that penicillin G is synthesized from three amino acids (L-α-
aminoadipic acid, L-cysteine, and L-valine) and the sidechain
precursor, phenylacetate. Furthermore, penicillin acyltrans-
ferase, an enzyme presumably involved in the final step of
penicillin G synthesis, catalyzes penicillin G synthesis from 6-
aminopenicillanic acid (6-APA) and phenylacetate. Therefore,
the effect of these substrates on penicillin production by the
immobilized mycelium was examined (Table II). Penicillin
productivity by immobilized mycelium in each substrate was
expressed relative to that observed in glucose-$(NH_4)_2SO_4$ medium.
As shown in Table II, the rate of penicillin production in a

Table II Effect of substrates on penicillin
          production

Substrates	Activity (%)
1 % Glucose, 0.5 % $(NH_4)_2SO_4$	100
0.05 % L-Cys, 0.05 % L-Val, 0.01 % DL- -Aminoadipate	150
0.01 % 6-APA	110

Each substrate mixture (in 0.05 M phosphate buffer,
pH 7.0) contained 0.01 % phenylacetate as well.

medium comprised of three amino acids was higher than that in a
6-APA medium. This might be attributable to transport barriers
for 6-APA across the cell membrane. However, as these amino
acids and 6-APA are expensive, glucose-$(NH_4)_2SO_4$ medium was
therefore employed for further work.

Penicillin Production by Immobilized Mycelium. Repeated
penicillin production by immobilized mycelium was examined in a
batch system. Experiments were performed as follows : about 1

g washed mycelium or 10 g immobilized mycelium (containing 1 g
wet mycelium) was incubated in a medium containing glucose for
5 h at 25°C. The amount of penicillin G produced was determined by bioassay. Native and the immobilized mycelia were
then washed thoroughly with 0.05 M phosphate buffer and stored
at 5°C. These experiments were repeated once a day. Penicillin
production by the washed mycelium decreased with repeated use,
as shown in Figure 1. On the other hand, the activity of the
immobilized mycelium increased initially and decreased gradually
thereafter. Activation of the immobilized mycelium was observed
initially. This might have been caused by growth of the mycelium
within the polyacrylamide gel or by a change in the permeability
of the mycelial cell wall. Then, with repeated use, the activity
decreased gradually. However, the estimated catalyst half-
life of the immobilized mycelium was six days, i.e., six times
longer than that of the washed mycelium. Some of the activity
loss in the immobilized preparation could stem from the inter-
mittent storage and testing procedure employed in these
experiments.

    The respiration activity of the immobilized mycelium was
examined by using an oxygen electrode. The rate of oxygen
uptake by the immobilized mycelium was about 30 % of that of
the washed mycelium. Furthermore, penicillin production by the
immobilized mycelium was performed under air and nitrogen
atmospheres and the penicillin produced was determined. Only
a small amount of penicillin was produced by the immobilized
mycelium under anaerobic conditions. Therefore, oxygen was
required for the system synthesizing penicillin in mycelium.
In addition, viable mycelium could be recovered from the
ground, immobilized mycelium in a germ-free box. These results
suggest that at least a fraction of the mycelium was living
after entrapment in the polyacrylamide gel and that the living
mycelium was responsible for production of penicillin from
glucose and $(NH_4)_2SO_4$.

Bacitracin Production by Immobilized Whole Cells

    The peptide antibiotic bacitracin is synthesized by the
protein thiotemplate mechanism present in Bacillus species
(14). Recently, cell free synthesis of bacitracin has been
also studied in some detail (14). Therefore, we selected
bacitracin for the production of secondary metabolite by immobilized whole cells.
    In this section, our studies on the production of bacitracin
by immobilized Bacillus sp. (KY 4515) from a simple nutrient
are described. Furthermore, the growth of the bacteria entrapped
in polyacrylamide gel was examined by electron microscopy.

    Immobilization of Whole Cells. Washed cells were re-
suspended in the starch-bouillon medium and incubated for 4 h

at 30°C. Bacitracin was produced by washed cells similar to the results described by Cornell and Snoke (15), and the rate of production was usually 1318 units/ml/h (cell concentration was about 20 mg-wet/ml). However, bacitracin productivity decreased with successive use of washed cells. Furthermore, collection of washed cells from reaction medium was difficult. Therefore, Bacillus sp. was immobilized in polyacrylamide gel.

The activity of bacitracin production was the highest when whole cells were immobilized in the gel prepared with 5 % total acrylamide (95 % of acrylamide monomer and 5 % BIS). On the other hand, no effect of BIS content on production of bacitracin was observed. However, the best productivity of bacitracin by immobilized whole cells was only 20-25 % of that by washed cells. These results suggest that the lower rate of bacitracin production is mainly caused by the inactivation of enzymes in whole cells with polymerization reagents (especially AA and APS) and may be partly due to hindered diffusion of the substrates and/or products through the gel.

Effect of Air, Media, Bacteria Content on Bacitracin Production. The effect of air (oxygen) on bacitracin production by immobilized whole cells was examined. The incubation of immobilized whole cells was carried out under aerobic conditions and under nitrogen. Bacitracin produced under anaerobic condition was only 30 % of that under aerobic condition.

Preliminary experiments indicated that the activity of bacitracin production by immobilized whole cells was remarkably reduced at the second reaction when a fermentation medium was used. Furthermore, the growth of leaked bacteria was observed when a fermentation medium (starch-bouillon) was employed for bacitracin production. Therefore, the effect of medium composition on bacitracin productivity was examined. As shown in Table III, bacitracin productivity in a medium

Table III   Effect of mediums on production
of bacitracin by immobilized
whole cells

	mediums	Bacitracin produced (U/ml)	
		1st	2nd[a]
Immobilized Whole cells	SB	12	4.1
	GB	7.2	2.0
	Peptone[b]	19	16
	Meat extract[c]	19	31

a) 2nd reaction was performed after 12 h storage at 4°C
b,c) Peptone and meat extract mediums contained 0.25 %
NaCl and 1×10^{-5} M $MnSO_4$.

containing meat extract or peptone was higher than that in a
medium containing carbohydrate. Furthermore, the productivity
was retained at the second reaction of immobilized whole cells.
This tendency was also observed in the case of bacitracin
production by washed cells. Therefore, 1 % peptone was selected
as the medium for repeated batch production of bacitracin.
Repeated batch production of bacitracin was attempted by using
immobilized whole cells and washed cells. The results are
shown in Figure 2. Bacitracin produced by washed cells decreased
gradually after several batch contacts. On the other hand,
bacitracin produced by the immobilized whole cells increased
with successive utilizations. The bacitracin produced in
saline appeared to result from synthesis from whole cell
nutrient pools.

Since bacitracin production in the 1 % peptone medium was
higher than that in the 0.5 % meat extract medium, the 1%
peptone medium (containing 0.25 % saline and 1×10^5 M $MnSO_4$)
was employed for further work. The growth of the leaked and
contaminating bacteria was not observed during the incubation
time.

The effect of the cell content in the gel on the bacitracin
production was examined. The rate of bacitracin production
increased with increasing the amount of whole cells in the gel.
A lag in bacitracin production was observed in the case of the
gel containing the smallest amount of bacteria. However, the
activity of each gel came to the same level with successive
uses.

Bacitracin Production from Amino Acids and ATP. Bacitracin
production by immobilized whole cells was attempted from
asparagine, aspartic acid, histidine, phenylalanine, isoleucine,
ornithine, lysine, glutamic acid, leucine, cysteine and ATP.
The results are shown in Figure 3. No activation of bacitracin
production was observed under these conditions. The amount of
bacitracin produced from amino acids and ATP was lower than
that from 1 % peptone.

The activity of bacitracin production increased with
increasing use cycles of immobilized whole cells and reached a
steady state maximum. Furthermore, apparent half-life of
bacitracin production was estimated to be at least one week.
On the other hand, the activation of bacitracin production was
not observed when amino acids and ATP were used as substrates.
In addition, the rate of activation increased with increasing
amount of bacteria in the gel. These results suggested that
the activation of bacitracin production was caused by the
growth of bacteria in the gel. Therefore, the gel was examined
by electron microscopy.

Microscopic Observation on Immobilized Cells. Electron
microscopic observation of the entrapped cells in the poly-

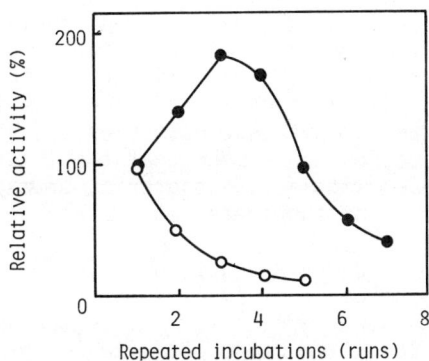

Figure 1. *Continuous production of penicillin by washed (○) and immobilized (●) mycelium*

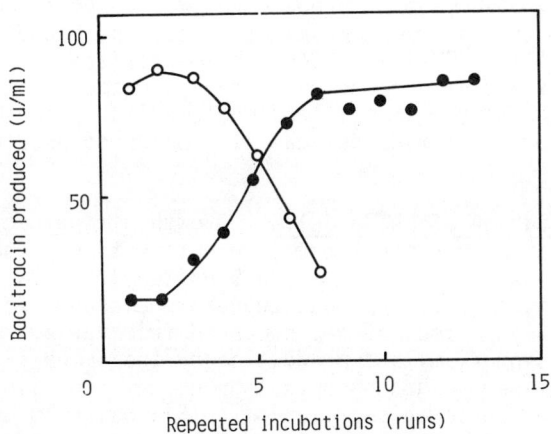

Figure 2. *Continuous production of bacitracin by washed (○) and immobilized (●) cells. The medium containing 1% peptone, 0.25% saline, and 1 × 10⁻⁵M MnSO₄ was used for the reaction.*

acrylamide gel was performed. Figure 4 shows the electron
micrographs of whole cells immediately after immobilization (a)
and also after 56 hours of use for bacitracin production (b).
The apparent growth of cells in the gel (especially at the
surface layer) was observed after successive uses (total 56 h
of reaction time) while the amount of cells evidently had
decreased in the center of the gel. The diffusion of oxygen
through the gel matrix may be a limiting factor of bacitracin
production. The growth of the bacteria during incubation
provides an obvious explanation for the increase in the activity
of bacitracin production with successive uses. A steady state
of bacitracin production could mean that the interstitial space
of the gel was filled with active bacteria, or may be controlled
by the diffusion rate of substrates and/or products through the
gel.
     The rate of bacitracin production by immobilized whole
cells was slightly higher than that of common fermentation.
     No attempt has been made as yet to use the immobilized
whole cells for a long time or to optimize the process for
antibiotic production.

α-Amylase Production by Immobilized Whole Cells

     Antibiotics such as penicillin G and bacitracin were
produced successfully by immobilized whole cells. No attempt
has previously been made to produce enzymes using immobilized
bacteria. In this section, α-amylase producing bacteria,
Bacillus subtilis FERM-P No. 2040 was immobilized in polyacrylamide gel and production of α-amylase was carried out in a
batch system using the immobilized whole cells. Growth of the
bacteria in polyacrylamide gel was also observed by electron
microscopy.

     Immobilization of Whole Cells. Growth of the bacteria
reached a stationary phase after 15 h cultivation. The maximum
rate of α-amylase production was 3000 units/ml/h. Bacillus
subtilis cultivated for 18 h was therefore employed for further
experiments. The effect of gel prepared from various acrylamide
concentrations on α-amylase production by the immobilized whole
cells was examined. The concentration of the crosslinking
reagent, N,N'-methylenebisacrylamide (BIS), was held constant
at 10 % (w/v) of the total acrylamide. The immobilized whole
cells prepared from 5% acrylamide showed the maximal production
activity of α-amylase. Therefore, 5 % total acrylamide concentration was employed for further tests.
     The effect of BIS content in acrylamide on the production
of α-amylase was examined. Total acrylamide concentration was
held constant at 5 %. As shown in Figure 5, the optimum BIS
content for the production of α-amylase was observed to be
above 15 %. The leakage of the bacteria from the polyacrylamide

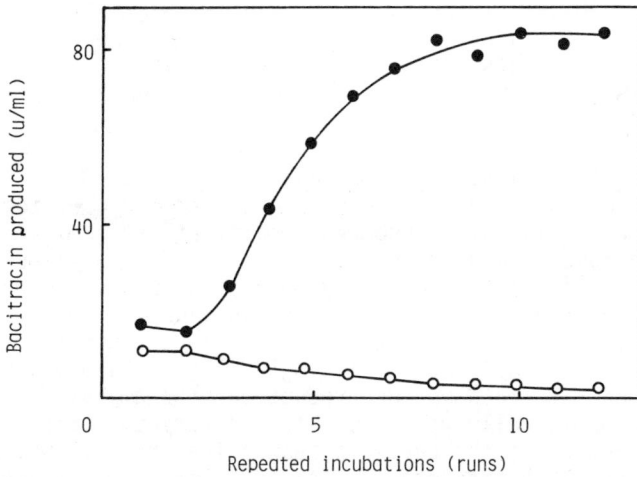

Figure 3.   *Bacitracin production from amino acids and ATP by immobilized cells: (○) amino acids (2.5mM) and ATP (2mM), (●) 1% peptone.*

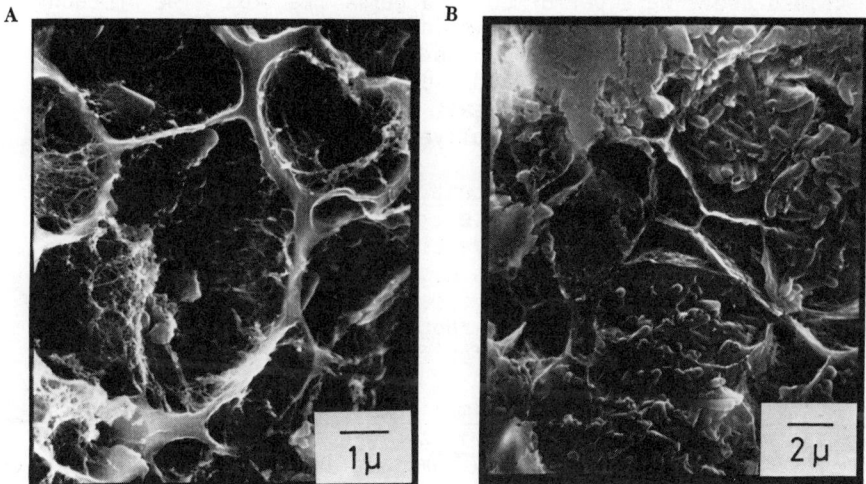

Figure 4.   *Electron micrographs of immobilized cells in the gel (bacitracin production): (a) gel immediately after immobilization; (b) gel after 56-hr incubation.*

gel described above was examined using the turbidimetric method.
 The gel was suspended in the reaction medium for 12 h at 30°C
and the turbidity at 660 nm was determined after incubation.
The leakage of bacteria was minimum when the bacteria were
immobilized in polyacrylamide gel prepared from acrylamide
containing 10-15 % BIS. Therefore, total acrylamide containing
85 % acrylamide and 15 % BIS was employed for immobilization of
the bacteria.

Effect of Nutrients on α-Amylase Production. Media
containing various natural nutrients are commonly used for the
production of α-amylase in submerged cultivation. However,
these media cannot be used in the production of α-amylase by
immobilized whole cells, because growth of contaminated bacteria
occurs during incubation. α-Amylase production by the immobilized whole cells was performed with various reaction media
containing 0.02 % $CaCl_2$ and 0.01 % $MgSO_4 \cdot 7H_2O$. Figure 6 shows
the effect of nutrients on the production of α-amylase. The
immobilized whole cells were incubated in various media for 5
h, because growth of the bacteria leaked from the gel was
observed after 12 h incubation. The activity of α-amylase in a
medium was determined by the blue value method. (16) Then the
gel was washed thoroughly with saline and immersed in 70 %
ethanol solution for 5 s (for sterilization). The gel was then
resuspended in the media and incubated for 5 h. All procedures
described above were performed in a germ-free box. The procedure
was repeated 12 times. As shown in Figure 6, the highest
activity of α-amylase was observed in the case of the medium
containing 1 % meat extract and 0.05 % yeast extract. Vitamins
and other trace nutrients in yeast extract may improve enzyme
formation. This medium was employed for further tests.

Effect of Whole Cells Content on α-Amylase Production.
The effect of the bacteria content in polyacrylamide gel on the
production of α-amylase was investigated. The reaction time
for the immobilized whole cells was 12 h (after one or two
cycles) and 5 h (after three cycles). The optimum content of
bacteria in the gel was in the range 5-7.5 %. The gel of 7.5 %
bacteria content was employed for further experiments.

Amylase Production at Optimal Conditions. Figure 7 shows α-
amylase production by the immobilized whole cells at the optimal
conditions. The production of α-amylase increased with
increasing reaction cycles. The activity of α-amylase in the
medium was 15000 units/ml at this time. The rate of α-amylase
production by the immobilized whole cells was almost the same
as that by fermentation described above.

Electron-Microscopic Observation of Immobilized Whole Cells.
The gel was examined under the scanning electron microscope.

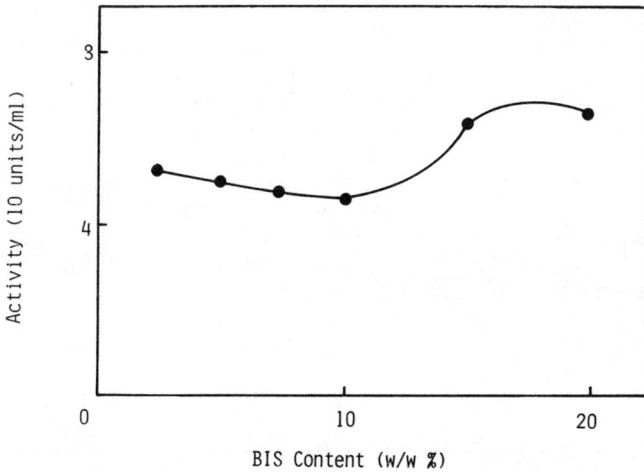

Figure 5.   Effect of BIS content in acrylamide on the production of α-amylase. The medium containing 1% meat extract, 0.02% $CaCl_2$, and 0.01% $MgSO_4 \cdot 7H_2O$ was used for the reaction.

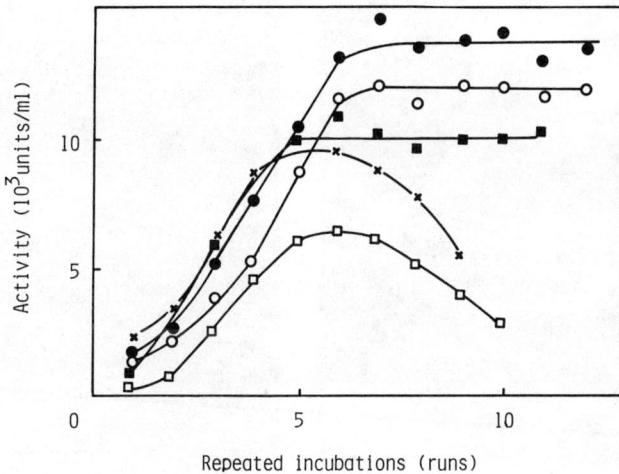

Figure 6.   Effect of nutrients on the production of α-amylase: (●) meat extract and yeast extract; (○) meat extract; (■) peptone and yeast extract; (□) peptone; (×) starch bouillon.

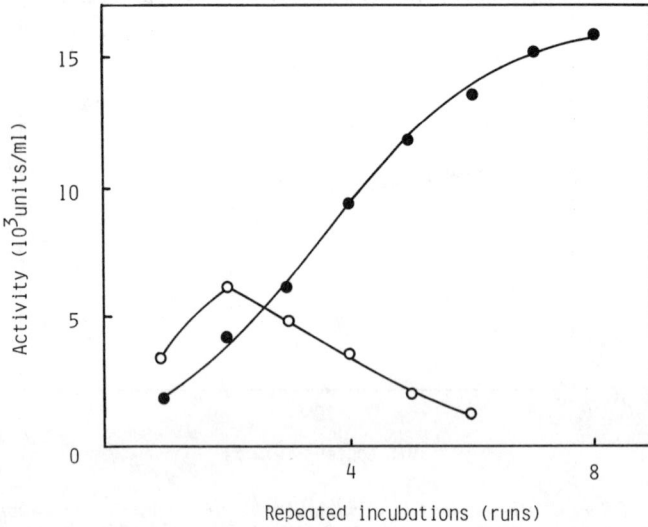

Figure 7.   Continuous production of α-amylase by immobilized cells at maximum conditions: (○) washed cells; (●) immobilized cells.

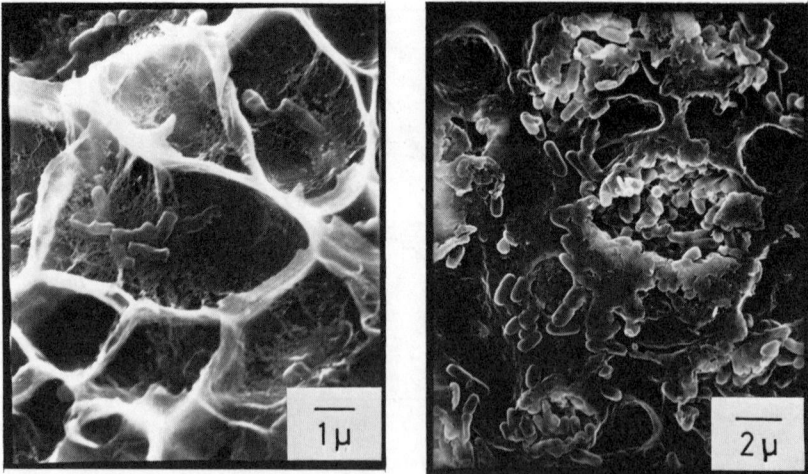

Figure 8.   Electron micrographs of immobilized cells in the gel (α-amylase production): (a) gel immediately after immobilization; (b) gel after 40-hr incubation.

Figure 8 (a) shows the electron micrograph of the gel
immediately after immobilization. Whole cells are well
distributed in the gel matrix. Figue 8 (b) shows the electron
micrograph of the gel after 40 h incubation in the medium. The
number of whole cells in the gel has clearly increased after
incubation in the reaction medium. Therefore, bacterial growth
occurred in the gel during incubation.

The growth of cells during incubation provides an obvious
explanation for the increase in the activity of α-amylase
production with successive utilization of immobilized whole
cells. The steady state of α-amylase production rate obtained
after seven cycles means that the interstitial space is filled
up with the active bacteria. Furthermore, electron-microscopic
observations suggest that the interstitial space of the polyacrylamide gel is large enough for the penetration of α-amylase.
No attempt has been made to use immobilized whole cells for a
long time for α-amylase production, and to optimize the process.
However, the results obtained in these preliminary experiments
indicate that the rate of α-amylase production by the immobilized bacteria was similar to that by fermentation.

Therefore, a column system using immobilized whole cells
appear s to have good potential for the industrial production
of enzymes in the future.

However, carriers with better mechanical properties and
better diffusibility must be considered in any future studies.

Literature Cited

1. Vieth,W.R., Wang, S., and Saini,R., Biotechnol.Bioeng., (1973)
   15, 565.
2. Mosbach, S. and Larson, P., Biotechnol.Bioeng., (1970) 12, 19.
3. Kierstan, M. and Buchke,C., Biotechnol.Bioeng., (1977), 19,
   387.
4. Chibata,I., Tosa,T., and Sato,T., Appl. Microbiol., (1974),
   27, 878.
5. Chibata,I., and Tosa,T., Adv. Appl. Microbiol., (1977) 22, 1.
6. Karube,I., Matsunaga,T., Tsuru,S., and Suzuki,S., Biochim.
   Biophys. Acta, (1976), 444, 338.
7. Karube,I., Matsunaga,T., Tsuru,S., and Suzuki,S., Biotechnol.
   Bioeng., (1977), 19, 1727.
8. Karube,I., Matsunaga,T., Mistuda,S., and Suzuki,S., Biotechnol.
   Bioeng., (1977), 19, 1535.
9. Karube,I., Mitsuda,S., Matsunaga,T., and Suzuki,S., J.
   Ferment. Technol., (1977) 19, 1535.
10. Karube,I., Matsunaga,T., and Suzuki,S., J.Solid Phase Biochem.
    (1977), 2, 97.
11. Matsunaga,T., Karube,I., and Suzuki,S., Anal.Chim.Acta, (1978)
    99, 233.
12. Slowinki,W., and Charm,S.E., Biotechnol.Bioeng., (1973) 15,
    973.

13. Halliday,W.J. and Arnstein, H.R.V., Biochem.J., (1956), 64,
    380.
14. Katz,E., and Demain,A.L.,Bacteriological Reviews, (1977)
    41, 449.
15. Cornell,N., and Snoke, J.E., Biochim.Biophys.Acta, (1964),
    91, 533.
16. Fuwa,H., J.Biochem., (1954) 41, 583.

RECEIVED March 29, 1979.

Denitrification and Removal of Heavy Metals from Waste Water by Immobilized Microorganisms

J. HOLLÓ and J. TÓTH
University of Technical Sciences, Budapest, 1521 Budapest, Hungary

R. P. TENGERDY and J. E. JOHNSON
Colorado State University, Fort Collins, CO 80523

This paper deals with the problem of removal of heavy metals and nitrate from waste waters by fixed-bed biological reactors. It is quite evident today that contamination by heavy metals increases simultaneously with the development of industry, e.g., the contamination of Lake Saint Claire near Detroit, the fish of the North Sea, the mercury poisoning in Minimata, Japan, or the ailments caused by cadmium in Itai-Itai. Nitrates and nitrites in drinking water may cause methhemoglobin anemia and may also contribute to the formation of nitrosamines which are highly suspected carcinogens. Nitrates and nitrites in effluents can cause eutrophication of the receiving bodies of water, the effusion of plants, seaweed and algae, and can increase BOD to such a level that capacity for self purification stops.

Biological denitrification and metal removal may be especially recommended as a treatment for bringing the level of these pollutants below the present stringent water quality standards; the removal of the last traces of these pollutants probably can be done more efficiently by biological than by physical or chemical means.

Biological denitrification

Many facultative heterotrophic bacteria (e.g., *Pseudomonas*, *Micrococcus*, *Denitrobacillus*, *Spirillium*, *Achromobacter*, etc.) are capable of denitrification. Denitrification takes place under anaerobic conditions, a precondition for the formation of a nitrate-reducing enzyme system.

Heterotrophic bacteria require the presence of an organic carbon source. Methanol proved to be the most suitable for giving maximal rate of nitrate reduction. The optimal methanol:nitrate ratio is 2.47 (g/g), which corresponds to a C:N mole ratio of 1 (1).

In the technology of denitrification with suspension cultures, the main problem is the removal of the suspended

0-8412-0508-6/79/47-106-073$05.00/0

bacteria by flocculation followed by sedimentation and/or
filtration. For the elimination of these problems, fixed-bed
reactors have been applied which have the additional advantage
of operating with high concentrations of microbial mass
($\underline{2},\underline{3},\underline{8}$). The other advantage of this method is that carrier
materials with relatively large specific surface areas increase
the adhesion and mass transfer in the heterogeneous phase ($\underline{3}$).
Many fixed-bed reactors are the extension of the trickling
filter sewage treatment process using various adsorbing
surfaces for microbes. Sikora and Keeney ($\underline{2}$) studied the
denitrification of a septic tank effluent by *Pseudomonas* at
various temperatures in continuous flow columns packed with
limestone chips using methanol as an energy source. Nearly
complete removal of nitrate was attained in 17 hours at 5°C,
13 hours at 13°C and less than 2 hours at 20°C; the kinetics
of the system was depicted as first order; an Arrhenius
relationship was shown with a calculated energy of 12-25
kcal/mole NO_3-N between 5-25°C. The optimum pH of *Pseudomonas
aeruginosa* is 7.0-8.2.

One problem with finely granulated inorganic carriers is
the high hydraulic resistance and hence decreased flow rate.
To avoid this problem plastics are now used rather extensively
as carriers for fixed-bed reactors ($\underline{9},\underline{10},\underline{12}$), including denitrification ($\underline{8}$). The shape and pore size of plastics can be
controlled to give desirable flow characteristics, and the
surface can be modified for better microbial attachment, e.g.
by plasma treatment. In earlier applications no rigorous
conditions or requirements existed for true immobilization of
the microbes. Consequently, many cells were detached from the
carrier to the effluent causing an increase in the BOD and COD
of the effluent water ($\underline{4}$).

Biological removal of heavy metals

The capacity of microorganisms, including algae, for
accumulation and metabolism of heavy metals is well docu-
mented. The applications range from sewage and industrial
waste treatment ($\underline{5},\underline{7}$) to ore leaching ($\underline{18}$), and to plutonium
removal from holding ponds ($\underline{17}$).

For similar reasons as mentioned in denitrification,
immobilized microorganisms would have advantages over suspended cultures in containing accumulated metals. In some
cases, as it will be shown below, heavy metal uptake and some
other microbiological functions, such as denitrification, can
be performed in one operation, using the same fixed-bed
bioreactor.

Heavy metal uptake is primarily based on the ability of
microbial surfaces to complex with metal cations. The negatively charged sugar units of polysaccharide chains, extending
from the microbial cell wall, may complex with metal cations.

Exocellular polyphosphate groups, excreted by bacteria, can also complex metal ions through chelation (7).   Membrane lipids complex lead and probably transuranics (14,15).

Results

Immobilization of *Pseudomonas aeruginosa*.   Seeking an improvement on some earlier fixed-bed biological reactors, the primary criterion in selecting a plastic carrier was the firm attachment of the microbe to the plastic.   This can be achieved if the microbe is either using the plastic as a carbon source and thereby burrows itself into the plastic, or if the plastic surface is treated, e.g., by plasma treatment, to facilitate chemical bonding.

After screening a wide range of plastic materials, the two most suitable ones were selected: polyvinyl chloride (PVC) films and melt blown polypropylene (PP) webs.   The softeners in PVC are the carbon source that facilitates firm embedment and continued metabolic activity in long duration continuous operations (20,21).   The melt blown PP has an unusually large surface which after plasma treatment has a large capacity for cell loading and very stable holding of bacteria, desirable for heavy metal removal.

Immobilization of *P. aeruginosa* on PVC.   A glass column of 3 X 150 cm was loosely packed with PVC sheets cut into small (less than 1 cm^2) pieces and sterilized with dry steam for 20 minutes.   The void volume was filled with a sterile mineral solution (MS) of: (NH_4NO_3, 0.1%; KH_2PO_4, 0.1%; Na_2HPO_4, 0.1%; NaCl, 0.1%; $MgSO_4 \cdot 7H_2O$, 0.1%; $CaCl_2 \cdot 2H_2O$, 0.01%; $FeCl_3 \cdot 6H_2O$, 0.01%.   A 24 hour beef-broth culture of *P. aeruginosa* (ATCC 13388) containing about 1.0 X 10^4 viable cells/ml was inoculated to the MS, 2 ml/l.   Temperature was kept constant at 25°C.   In Figure 1 it can be seen that the change in viable cell numbers was rather slow, developing in the column reactor on the plastic surface in an order of magnitude of 10^7 cells/cm^2 in 14 days.   If an additional C-source, glucose or methanol, was added, growth accelerated. It follows from the figure that the plastic may serve as sole carbon source required for the multiplication of bacteria but an additional carbon source promotes the initial phase of attachment, due to the better adhesion of a large number of quickly multiplying bacteria, which become attached as a new layer to the layer of bacteria fixed on the plastic surface. The same holds true for the addition of various trace elements as nutrient supplements, e.g., calcium, magnesium, iron, manganese and phosphate salts, which not only stimulate the multiplication of cells, but by surface charge alterations also affect physical attachment between negatively charged bacteria and positively charged surface groups of the plastic

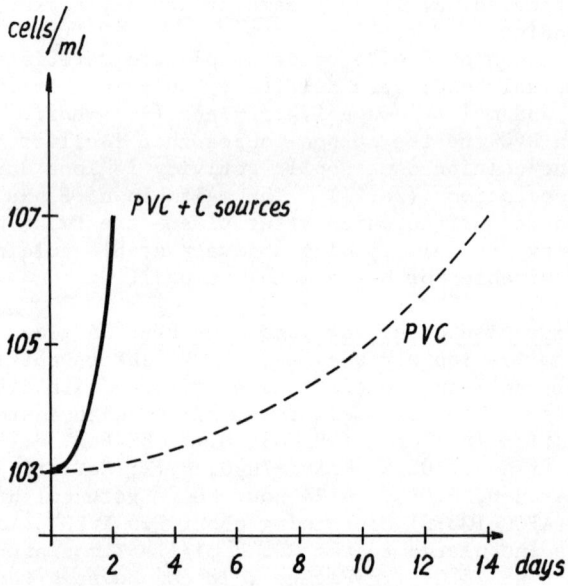

Figure 1. *Growth of* Pseudomonas aeruginosa *on PVC with or without added carbon source*

(6). Scanning electron micrographs (Figures 2 and 3) of the immobilized bacteria clearly show the mass of rod-like viable bacteria in the process of cell-division on the surface of the plastic. This proves immobilization of a considerable amount of bacteria on the soft PVC surface as they have not become detached from the surface even during preliminary treatment required for electron microscopy. No bacteria could be detected, on the other hand, on the surface of less suitable plastic beds.

The total number of bacteria attached to the plastic surface was determined by the Lowry test (16), the number of viable bacteria was determined colorimetrically by the triphenyl-tetrazolium-chloride (TTC) reduction test (13).

Immobilization of *P. aeruginosa* on PP. Melt blown, fine PP filaments (<1 μm in diameter) are inert, but large surface area carriers, allowing a high cell loading density (up to a maximum of 1 g cells/1 g plastic). The efficiency of the PP web for bacterial attachment can be greatly enhanced by O_2 plasma treatment (wetting) of the surface.

The PP web was plasma treated in a sealed glass tube glow discharge apparatus operating at 13.56 MHz frequency, 8 watt output, 1.5 mm Hg O_2 pressure. A few seconds (2-3) treatment caused optimal surface wettability and increased charge and thus better attachment. Longer treatments burned the surface of the plastic.

P. aeruginosa cells were immobilized on plastic surfaces by immersing the plastics into a growing *P. aeruginosa* culture for 72 hours. The attachment is basically a firm adhesion process in which acidic polysaccharide fibrils connect the bacteria to the surface which is coated and wetted by a bacterial extracellular protein (6). Sloughing of the bacteria from the surface, in our experience, was minimal.

Denitrification. The PVC-bacteria columns prepared as described above have been used for studying the effect of flow rate and input nitrogen concentration on denitrification. For better utilization of the volume of the column and in order to increase the capacity of denitrification, methanol was applied as an additional carbon source (in a ratio of C/N = 1). Input was on top of the column and output was regulated through an overflow device (Figure 4). In this system, the immobilized bacteria acting as a large surface bioadsorbent, the floating particles (cells, cell debris, others) are effectively retained. Gas outlet was at the top through a liquid trap in order to ensure anaerobic conditions. Input and output nitrate concentrations were determined by standard water quality control procedures: after evaporation by salicylic acid and dissolution with sulfuric acid, the yellow salicylic acid nitrate complex obtained upon alkalization was determined

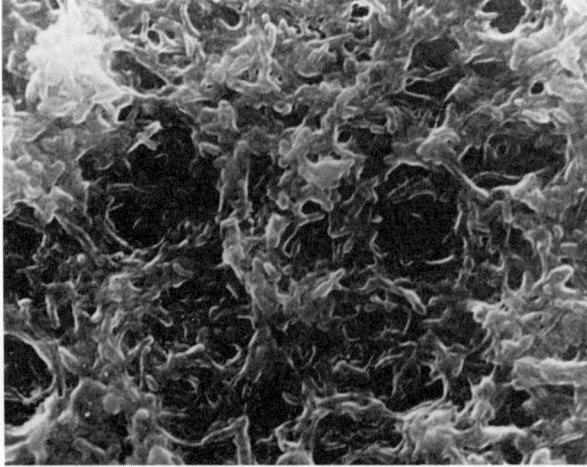

Figure 2.   *Electromicrograph of* Pseudomonas aeruginosa *embedded in a PVC sheet*

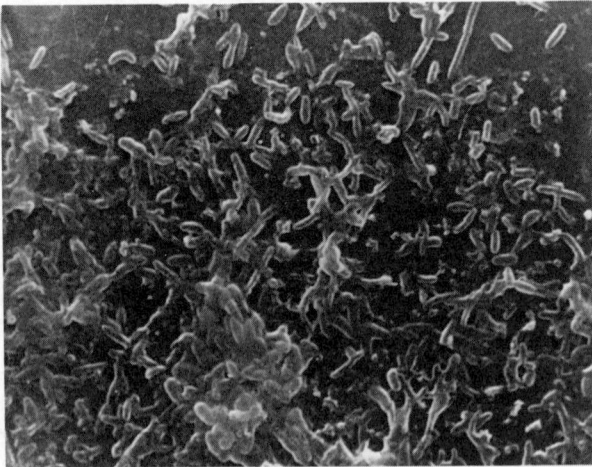

Figure 3.   *Electromicrograph of* Pseudomonas aeruginosa *embedded in a PVC sheet*

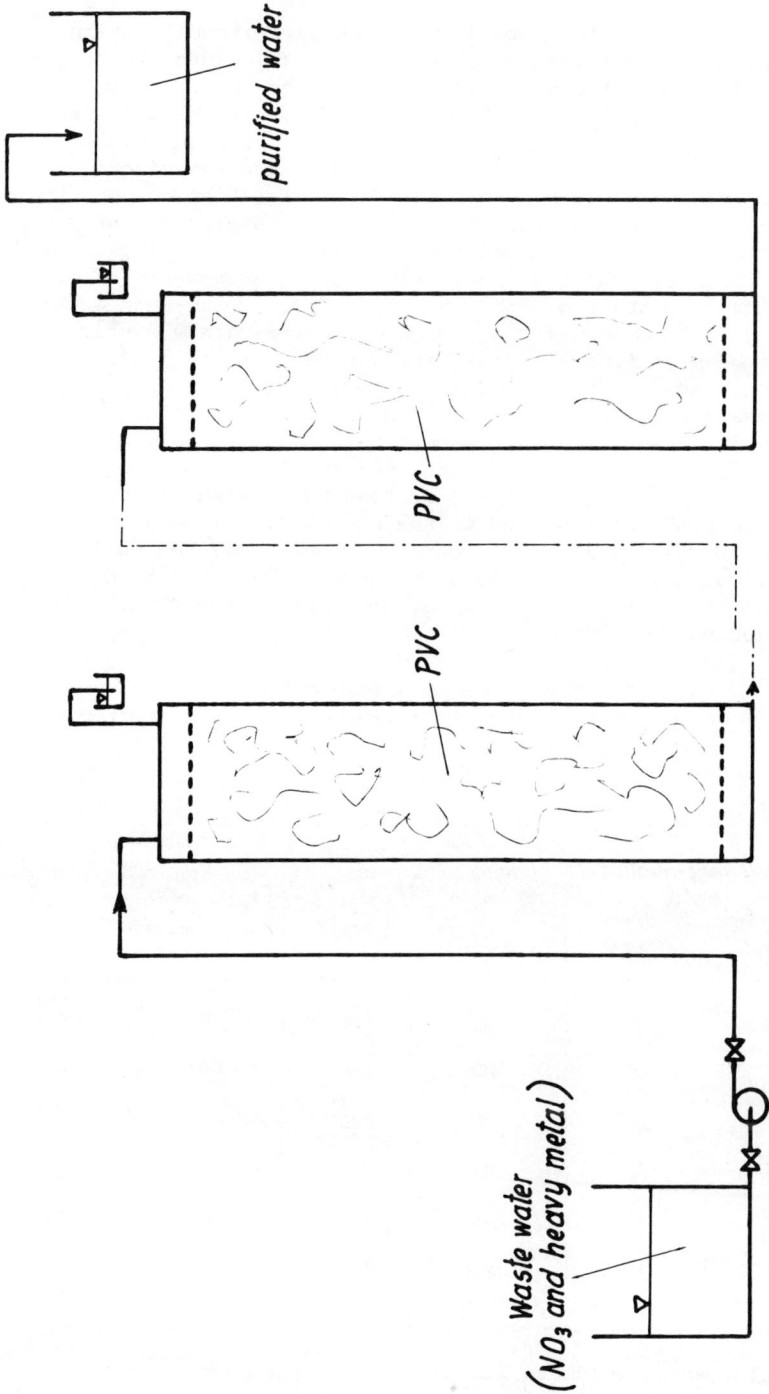

Figure 4. Flow diagram of the PVC bioreactor

photometrically. In spite of the effective immobilization
technique and optimization of the flow rate, eluted bacteria
(10^2-10^3 cells/ml) could always be detected in the effluent
water. For trapping these bacteria a freshly filled column
was attached to the end of the system. After saturation of
this new column with bacteria, the old column was discon-
nected, the new column became the denitrification column, and
a fresh filter column was attached. By this arrangement the
operation can be maintained continuously.

Table I shows average values for measurements of denitrification capacity and efficiency on various columns in experiments carried out during a period of approximately 2 years.
As can be seen from the data, the increase in the denitrification capacity of each column (3.6-24.3 kg nitrate/m^3 per
day) was nearly proportional to the increase in feeding rate
and in daily load. At the same time, the efficiency of
denitrification has dropped from 84% to 71% as feeding rate
increased. With the increase of feeding, a simultaneous
increase could be observed in the number of cells on the
column as well as in the eluent. The data measured on the
individual columns prompted us to apply serially linked
columns by which nitrate removal could be achieved with 99%
efficiency.

Table I. Denitrification capacity and efficiency
of column reactors using immobilized P. aeruginosa

Nitrate concentration [mg/l]		Feeding speed [ml/h]	Load [g NO₃/day]	Capacity [kg reduced NO₃/day m³ volume]	Efficiency [%]
Influent	Effluent				
106	17	427	1·08	3·6	83·8
114	18	488	1·32	4·5	84
116	19 5	727	2·03	6·6	83·1
115	22	1334	3·7	11·7	81
119	30	2115	5·95	17·5	75
119	34	3093	8·6	24·3	71

Useful reactor volume: 250 ml

Figure 5 shows the correlation between capacity, efficiency and nitrate loads. Although there are some deviant
points among the data plotted, when linear regression was
performed on the basis of individual data, the correlation
coefficient was found to be very good for capacity (0.99) and
satisfactory for efficiency (0.7). This verifies the correct-
ness of the relationships and the straight lines drawn. The
intersection of the 2 lines has been regarded as an operational parameter of the individual columns, thus in our
further experimental design a denitrification efficiency of
75-80% and a capacity of 12-13 kg nitrate/m^3/day have been
calculated.

Simultaneous removal of heavy metals and nitrate. In
addition to denitrification, the above series of columns were
also used for the removal of heavy metals. A decrease in the
denitrification capacity of the column was used as an index of
the critical heavy metal concentrations causing toxicity to
the *P. aeruginosa*. The toxic effect of metals (lead, chromium, copper, cadmium and zinc) was examined in a separate
study (unpublished data). Of these metals copper proved to
have the highest toxicity as it fully prevented denitrification in a very short time at a concentration of 10 ppm.
Regarding their toxic effect the next in order was lead,
followed by cadmium, zinc, and chromium.
The removal of heavy metals on fixed-bed columns was
performed primarily with lead and zinc in a concentration of 5
ppm as this proved to be most critical in the practical
realization of large scale operations. The metal content of
input and output waste waters was determined by atomic absorption spectrophotometry. In the series of experiments the
appropriately treated columns were kept in operation only for
a certain period of time (3-400 hours), when denitrification
capacity suddenly dropped in a few hours from 90-95% to 0,
indicating that the microbes accumulated a lethal dose of the
toxic heavy metal (pH = 7.5, at 25°C, with continuous feedings
at the 5 ppm level). Cell death was also verified by the lack
of TTC reduction.
The equipment used for the recycling of waste water in a
Hungarian chemical plant has been designed on the basis of the
above experimental results. Contamination in the waste water
of this plant exceeded the permissible limit for zinc, chromium, barium, aluminum, nickel, lead and nitrate. Based on
above experiments, the metal ions present in toxic concentrations were first precipitated by the addition of lime
hydrate and sedimented as hydroxide in the pH range: 8.5-9.0
(5). After readjustment to pH 7.5, the very diluted metal
ions (about 1 ppm) in the supernate were transferred into a
biological reactor and then removed with or without nitrate by

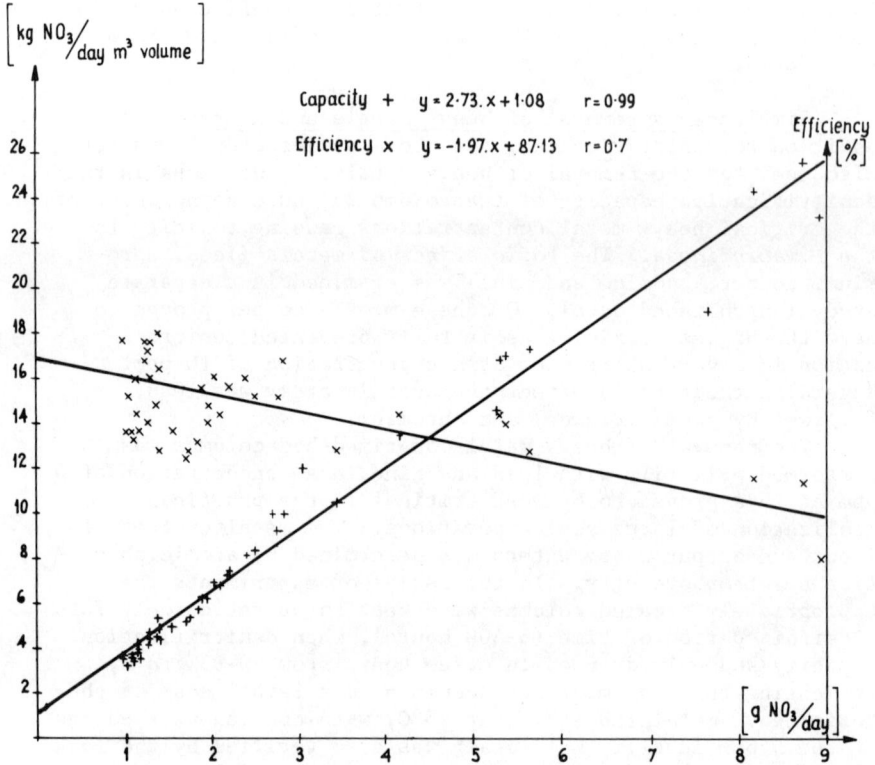

Figure 5.   Correlation between nitrate removal capacity and efficiency

means of *Pseudomonas* cells immobilized on a PVC bed as described above. The direction of the flow in the biological
reactor was changed weekly in order to improve the efficiency
of filtering. Filtration of possible eluted microbes was
carried out on a sand-filter, which assured microbe-free
effluent water. After regeneration by flushing with water,
the regenerate was refed to the system in the first step of
the process. The water thus purified was suitable for recycling and provided part of the water supply of the plant for
technological purposes.

Biological plutonium removal. *P. aeruginosa* cells
attached to PP were used efficiently for the removal of
plutonium from aqueous systems. Pu removal experiments were
conducted in small scale batch or continuous column operations.

In the batch experiment 0.1 g squares of the bioadsorbent
were immersed in $PuCl_4$ solutions or PuO_2 suspensions for 6
hours. A 0.1 g sample of the PP bioadsorbent was able to
remove 96% of 1.7 nCi $PuCl_4$ activity (Table II). The removal
of PuO_2 depended on particle size. From 1 µm diameter mono-
disperse particles 34% was removed, from 0.59 µm and 0.13 µm
particles 13% and 12% respectively was removed under the same
conditions as for $PuCl_4$ (Table III). This finding suggests
that the insoluble PuO_2 particles are merely entrapped in the
filaments, rather than adsorbed to the bioadsorbent; smaller
particles escaped entrapment.

In the continuous Pu removal experiments small 10 X 100
mm glass columns were packed with PP bioadsorbent (cell
loading: 0.3 g cells/g plastic; column load: 1.5 g bioadsorbent). This column removed 75-80% of 1.7 nCi/100 ml
$PuCl_4$ activity at a 15 ml/hour flow rate in a 12 hour operation. Since the maximal Pu removal capacity of this bioadsorbent was estimated as about 600 nCi/g cells, such a
column can be operated at the 1-5 nCi/100 ml level of activity
theoretically for at least a month before exhaustion. For
practical reasons, due to decreasing viability caused by the
toxicity of other heavy metals, not Pu, present, columns
should be replaced every week or two for optimal operation.
If a series of columns were operated in a counter current
fashion, the first column in the series being replaced weekly
with a fresh one, a 99-100% removal of Pu activity might be
achieved.

Although many synthetic adsorbents could remove Pu as
efficiently as bacteria, the economy and renewability of this
bioadsorbent makes it particularly attractive for removal of
trace amounts of Pu or other transuranics from waste waters.
The use of cheap plastics as carriers eliminates the need for
regeneration of the adsorbent; the Pu concentrated on the
bioadsorbent can be safely disposed.

Table II. PuCl$_4$ uptake by *Pseudomonas aeruginosa*
immobilized on plasma treated polypropylene web

Sample	Mean Pu uptake %
Control	6.7 ± 2
Immobilized cells	96 ± 15

PuCl$_4$ activity: 1.7 nCi

Sample size: 0.1 g

Cell loading: 0.3 g/g polypropylene

Table III. PuO$_2$ uptake by *Pseudomonas aeruginosa*
immobilized on polypropylene web

PuO$_2$ particle size (μm)	Mean uptake %
1.0	34 ± 9
0.59	13 ± 8
0.13	12 ± 1.5

PuO$_2$ activity: 1.5 nCi

Conclusions

Fixed-bed bioreactors, using immobilized *P. aeruginosa*
cells are suitable for the simultaneous or separate removal of
nitrates and heavy metals from very diluted solutions. The
availability of a large amount of plastic waste makes it
attractive as a cheap carrier for waste removal. The choice
of plastic depends on the intended application: for denitrification PVC is adequate (maximum viability of cells but less
stability), for Pu removal PP is better (maximum stability,
lessened viability). The plastic based bioreactors are highly
competitive with others both in performance and in economy.

Literature Cited

1. Ferris, J. S., Owens, R. W. Pilot scale, high-rate
    biological denitrification. J. Water Poll. Control Fed.
    47(8):2043 (1975).
2. Sikora, L. J., Keeney, D. R. Denitrification of nitri-
    fied septic tank effluent. J. Water Poll. Control Fed.
    48(8):2018 (1976).
3. English, J. W., Carry, Ch. W., Masse, A. N., Pitkin, J.
    B. and Dryden, F. D. Denitrification in granular carbon
    and sand columns. J. Water Poll. Control Fed. *46*(1):28
    (1974).
4. Bailey, D. A. and Thomas, E. V. The removal of inorganic
    nitrogen from sewage effluents by biological denitrifi-
    cation. Wat. Poll. Cont. *74*(5):497 (1975).
5. Lanouette, K. H. Heavy metals removal. *In* Chemical
    Engineering (Desk-book Issue), October 17, p. 73-80
    (1977).
6. Marshall, K. C. Mechanism of adhesion of marine bacteria
    to surfaces. IAEA/ERDA Symposium on Transuranium Nuclides
    in the Environment. San Francisco, California, November
    17-21 (1975).
7. Dugan, P. R., *et al*. Metal binding by microorganisms.
    In Advances in Water Pollution Research, Vol. 2. S. H.
    Jenkins (ed.), Pergamon Press, New York, pp. 111-120
    (1970).
8. Requa, D. A. and Schroeder, E. D. Kinetics of packed bed
    denitrification. J. Water Poll. Control. Fed. *45*:1696
    (1973).
9. Quirk, T. P. Scale up and process design techniques for
    fixed film biological reactors. Water Res. (G. B.)
    6:1333 (1972).
10. Pirt, S. J. Quantitative theory of the action of mi-
    crobes attached to a packed column. J. Appl. Chem.
    Biochem. (G. B.) *23*:398 (1973).
11. Mohan, R. R. and Li, N. N. Nitrate and nitrite reduction
    by liquid membrane encapsulated whole cells. Biotechn.
    and Bioeng. *17*:1137 (1975).
12. Bunch, R. L. Biological Filters: literature review. J.
    Water Poll. Control Fed. *46*:1121 (1974).
13. Tengerdy, R. P., Nagy, J. G. and Martin, B. Quantitative
    measurement of bacterial growth by the reduction of
    tetrazolium salts. Appl. Microbiol. *15*:954 (1967).
14. Tornabene, T. G. and Edwards, H. W. Microbial uptake of
    lead. Science *176*:1334 (1972).
15. Tornabene, T. G. and Peterson, S. L. Interaction of lead
    and bacterial lipis. Appl. Microbiol. *29*:680 (1975).
16. Lowry, O. H., *et al*. Protein measurement with the Folin
    phenol reagent. J. Biol. Chem. *193*:262 (1951).

17. Johnson, J. E.   Study of plutonium in aquatic systems of
     the Rocky Flats environs.   Final Techn. Report, Colorado
     State University, Fort Collins (1974).
18. Schwartz, W. (ed.).   Conference, Bacterial Leaching,
     Verlag Chemie Internat., New York (1977).
19. Task group on metal accumulation.   Environ. Physiol.
     Biochem.   *3*:65 (1973).
20. Alexander, J.   Nonbiodegradable and other recalcitrant
     molecules.   Biotechn. Bioeng.   *15*:611 (1973).
21. Stahl, W. H. and Ressen, H.   The microbiological degra-
     dation of plasticizers.   Appl. Microbiol.   *1*:30 (1953).

RECEIVED March 29, 1979.

Synthesis of Coenzyme A by Immobilized Bacterial Cells

SAKAYU SHIMIZU, YOSHIKI TANI, and HIDEAKI YAMADA

Department of Agricultural Chemistry, Kyoto University, Kyoto 606, Japan

A variety of enzymes have been immobilized with solid supports, and applications in industrial and clinical fields have been discussed. Immobilization of whole microbial cells as a bag of enzymes also has been developed by a number of workers. Current applications of continuous enzyme reactions using immobilized microbial cells for the transformation of useful compounds are mainly carried out by the action of a single enzyme. However, many useful compounds are usually produced, especially in fermentative processes, by the action of several kinds of enzymes. Only a few papers have been published on the immobilization of microbial cells operating in multi-step enzyme reactions, and little is known about the conditions for immobilization and the enzymatic properties of the immobilized cells containing the multi-step system (1,2).

CoA plays an important role in various metabolic reactions. This compound is used widely in biochemistry, and its usefulness in medicine and in analytical fields has been discussed. The recent discovery of an acyl carrier protein containing an intermediate of CoA biosynthesis, P-pantetheine, as a prosthetic group, has emphasized the significance of CoA and its derivatives in biochemistry. The biosynthesis of this coenzyme from pantothenic acid, cysteine and ATP requires five sequential enzymatic steps as follows (3):

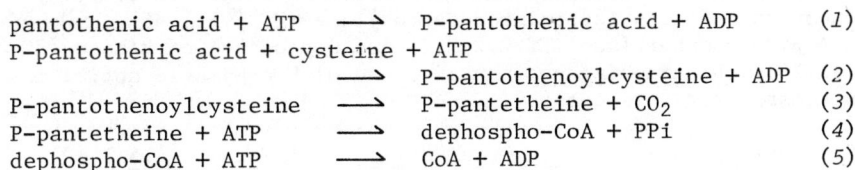

$$
\begin{array}{lll}
\text{pantothenic acid} + \text{ATP} \longrightarrow \text{P-pantothenic acid} + \text{ADP} & (1) \\
\text{P-pantothenic acid} + \text{cysteine} + \text{ATP} & \\
\qquad \longrightarrow \text{P-pantothenoylcysteine} + \text{ADP} & (2) \\
\text{P-pantothenoylcysteine} \longrightarrow \text{P-pantetheine} + CO_2 & (3) \\
\text{P-pantetheine} + \text{ATP} \longrightarrow \text{dephospho-CoA} + \text{PPi} & (4) \\
\text{dephospho-CoA} + \text{ATP} \longrightarrow \text{CoA} + \text{ADP} & (5) \\
\end{array}
$$

We have attempted to synthesize this coenzyme using microbial cells as enzymes. After extensive screening we found that Brevibacterium ammoniagenes is suitable for this purpose (4,5,6). With

0-8412-0508-6/79/47-106-087$05.00/0
© 1979 American Chemical Society

this bacterium we established a new and simple process for the
large scale production of CoA (6-12). The immobilization of this
whole system for CoA biosynthesis is an important example of a
multi-step enzyme reaction. We here summarize our studies on the
production of CoA by microorganisms focusing on the process with
immobilized cells.

Principle of CoA Synthesis and Selection of Suitable Microorganisms (5,7,12)

In the biosynthetic pathway of CoA from pantothenic acid and
cysteine, four moles of ATP are required for one mole of CoA (see
equations 1-5 described in the previous section). We previously
reported that adenosine or AMP is phosphorylated to ATP in a high
yield during the fermentative process of yeasts, when high concentrations of inorganic phosphates and glucose are present (13).
Thus, it should be possible to synthesize CoA by coupling the
microbial CoA synthetic system with an ATP-generating system as
illustrated in Fig. 1. About 300 strains of microorganisms were
randomly screened with an assay mixture composed of glucose, AMP,
potassium phosphate, sodium pantothenate, cysteine, magnessium
sulfate and dried cells of each microorganism as enzyme. Only a
few strains of yeast accumulated small amount of CoA (20-200 µg/ml).
Many yeasts could not accumulate CoA inspite of their strong
activity for ATP generation. When ATP was added to the reaction
mixture in place of AMP, several bacterial strains belonging to
the genera Sarcina, Corynebacterium, Micrococcus, Brevibacterium
and Nocardia produced 200-800 µg of CoA per ml. Of these, B.
ammoniagenes IFO 12071 had the most remarkable accumulation of CoA.
We used this bacterium to investigate various conditions for CoA
production. Based on our results, three different methods for
production were developed; the cell method, the fermentative
method and the immobilized cell method. All essentially are based
on the same principle, that CoA is enzymatically synthesized from
pantothenic acid, cysteine and ATP by cells of B. ammoniagenes,
though the processes differ. The reaction proceeds as described
above. Reactions 1, 2, 3, 4 and 5 described in the previous
section are respectively catalyzed by pantothenate kinase (EC 2.7.
1.33), phosphopantothenoylcysteine synthetase (EC 6.3.2.5),
phosphopantothenoylcysteine decarboxylase (EC 4.1.1.36), dephospho-
CoA pyrophosphorylase (EC 2.7.7.3) and dephospho-CoA kinase (EC 2.
7.1.24). B. ammoniagenes contains the high enzymatic activities
necessary for this pathway.

Synthesis of CoA by Free, Dried Cells (Cell Method, 4,5,7,12)

A typical time course for CoA synthesis by this method is
shown in Fig. 2. The amounts of CoA synthesized from pantothenic
acid and from pantetheine were about 1 mg/ml and 1.3 mg/ml, respectively. An addition of a surfactant, sodium laurylbenzene-

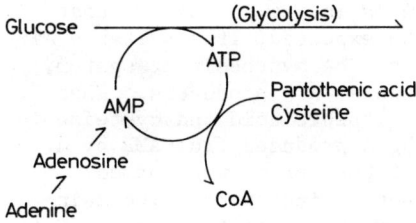

Figure 1.   Synthesis of CoA from pantothenic acid and cysteine coupled with
ATP-generating system

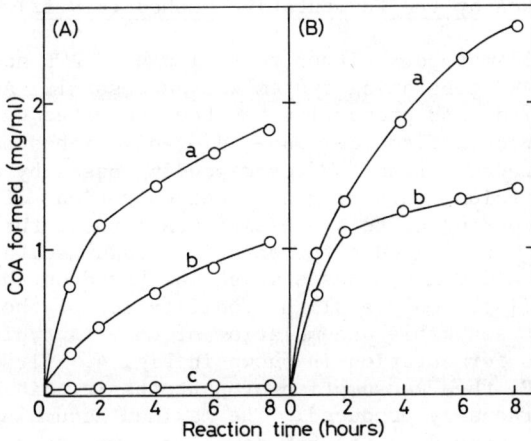

Figure 2.   Time course for CoA synthesis by dried cells of B. ammoniagenes.
(A) Synthesis from pantothenic acid: reaction mixture (1 mL) containing 5 μmol sodium
pantothenate, 10 μmol cysteine, 15 μmol ATP, 10 μmol magnesium sulfate, 150 μmol
potassium phosphate buffer, pH 6.0, and 100 mg dried cells of B. ammoniagenes was
incubated at 37°C with (a) or without (b) 2 mg of sodium laurylbenzenesulfonate. A
mixture without sodium pantothenate (c) was used as a control run. (B) Synthesis from
pantetheine: the reaction conditions were the same as those in (A) except that an equi-
molar amount of pantethine was used in place of sodium pantothenate.

sulfonate, brought about an increse in the amounts of CoA synthesized. Removal of the permeability barrier seems to be necessary
for substrates to contact with the enzymes; therefore, cells are
usually dried or treated with a surfactant. This was confirmed
by the leakage of enzymes necessary for CoA biosynthesis from the
cells when treated with the surfactant. This suggests that CoA
synthesis occurs, at least in part, extra-cellularly. Another
interesting feature of this process is that the ATP used for CoA
synthesis is not regenerated. Results in Fig. 3 suggest that an
increased accumulation of CoA may be expected, if a suitable ATP-
generating system is coupled with the CoA synthesizing system;
thus, we tested several yeast strains as ATP producers. When B.
ammoniagenes was incubated with pantothenic acid and cysteine in a
reaction mixture in which ATP had been produced from AMP by a
yeast Hansenula jadinii, 0.5-1 mg of CoA per ml was formed.
However, this amount of CoA could not be increased. The main
difficulty was to maintain an optimum ATP concentration for CoA
formation.

Production of CoA by the Fermentative Method (8,9,12)

     In the cell method, neither recycling used ATP nor supplying
it from other ATP-generating system was successful. Another
useful process for ATP production has been reported by Tanaka et
al. (14), in which nucleoside mono-, di- and triphosphates were
successfully produced from the corresponding bases by cultivating
bacteria in a medium containing high concentrations of glucose and
salts. This prompted us to synthesize CoA through the growth of
the bacterium. We devised a fermentation medium suitable for both
ATP generation and CoA synthesis which was based on the results of
Tanaka et al. (14). Only a slight modification of their orignianl
medium caused a remarkable accumulation of CoA. A typical time
course for this fermentation is shown in Fig. 4. First, phosphorylation of AMP, then consumption of pantothenic acid occurred,
then CoA was gradually produced. The maximum accumulation, about
3.5 mg/ml, was reached on the 6th day. In some cases, values of
more than 5 mg/ml were obtained. Because of vigorous aeration
during cultivation, CoA in the culture broth was present as the
disulfide. Production of CoA with adenine or adenosine instead of
AMP was also possible (yield: 1.5-2.5 mg/ml).

Synthesis of CoA by the Immobilized Cell Method (2)

     First, we tested various immobilization methods using cells
of B. ammoniagenes as the test organism. Entrapment of cells in a
cellophane tube, in polyacrylamide gel and in polyvinylalcohol
film gave good results. Immobilization with ethylene maleic an-
hydride, glutalaldehyde and polysilastic resin produced inactive
cells. Cells in the cellophane tube were active after seven 10
hr-incubations. During this period more than 80 mg of CoA was

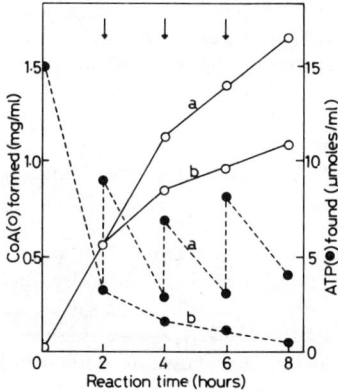

Figure 3.   Effect of ATP feeding on CoA synthesis. The reaction mixture (1 mL) containing 10 μmol sodium pantothenate, 10 μmol cysteine, 15 μmol ATP, 10 μmol magnesium sulfate, 150 μmol potassium phosphate buffer, pH 6.0, and 100 mg dried cells of B. ammoniagenes was incubated at 37°C. ATP (5 μmol each) was added to the mixture at the time indicated by arrows (a). A mixture without ATP feeding was used as a control run (b).

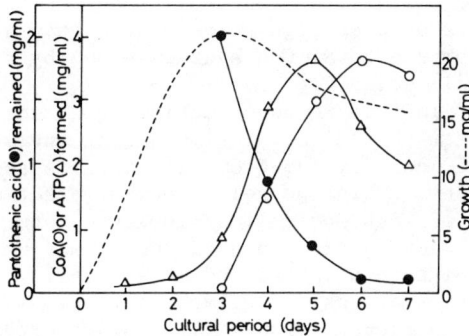

Figure 4.   Time course of CoA production during cultivation of B. ammoniagenes. B. ammoniagenes was cultivated for 24 hr at 28°C in a seed medium composed of glucose (1%), peptone (1.5%), yeast extract (0.1%), K₂HPO₄ (0.3%), NaCl (0.2%), and MgSO₄ · 7H₂O (0.02%), pH 7.0. The grown seed (10%) was inoculated in a fermentation medium composed of glucose (10%), urea (0.6%, autoclaved separately), peptone (0.8%), yeast extract (0.5%), K₂HPO₄ (2%), MgSO₄ · 7H₂O (1%), biotin (0.0003%), and AMP (0.2%), pH 7.6. Cultivation was run at 28°C with vigorous aeration. To the 3-day culture, calcium pantothenate (0.2%), cysteine (0.2%), and cetylpyridinium chloride (0.1%) were added, and the cultivation was continued for another 4 days.

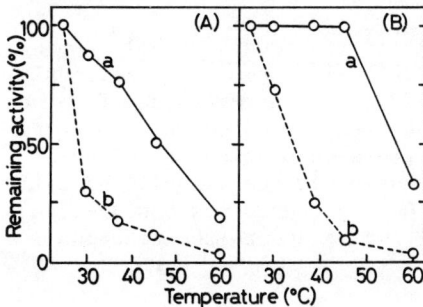

Figure 5.   Effect of temperature on the stability of immobilized cells. The gel-entrapped dried cells (a) were incubated in 0.01M potassium phosphate buffer, pH 7.0, for 3 hr at various temperatures, as indicated. After filtration, each gel (1.88 g) was incubated with 20 μmol sodium pantothenate, 100 μmol cysteine, 150 μmol ATP, 100 μmol magnesium sulfate, 1500 μmol potassium phosphate buffer, pH 6.5, 20 μmol CTP, and 10 mg sodium lauryl-sulfate in a total volume of 10 mL. The reaction was carried out at 37°C for 5 hr with shaking. Free, dried cells (b) were used as control. (A) Accumulation of CoA; (B) consumption of pantothenate.

synthesized and loss of the initial activity of the cells was
about 50%. The same amount of free, dried cells synthesized 25 mg
of CoA in the first incubation, after which the cells became inactive. The cells trapped in polyacrylamide gel and in polyvinylalcohol film also withstood repeated incubations, although their
initial activities were somewhat lower than those of free, dried
cells. We investigated various conditions for CoA synthesis with
polyacrylamide gel-entrapped cells.

General Properties of Immobilized Cells (2). Cells were
trapped in polyacrylamide gel by the procedure described by Chibata
et al. (15). The most active gel was obtained when 15-30% acrylamide monomer was polymerized at 0°C with 0.8% N,N-methylenebis-
acrylamide, 0.5% N,N,N',N'-tetramethylethylenediamine and 15-20%
dried cells.

The gel was stable for at least 45 days when stored at 0°C in
0.01 M potassium phosphate buffer, pH 7.5. Under the same conditions free, dried cells were completely inactivated within a week.
The gel was more stable at high temperatures than were free, dried
cells as judged by the total CoA synthesis and by the phosphorylation of pantothenic acid (Fig. 5).

As described in the previous section, the addition of a surfactant produced remarkable stimulation during the synthesis of CoA.
In the immobilized cell system, some surfactants showed similar
stimulative effects. When the gel in which intact cells were
trapped was treated with sodium laurylsulfate before the reaction,
it was remarkably activated. An addition of sodium laurylsulfate
to the reaction mixture with trapped intact cells caused an increased accumulation of CoA. Similar observations have been
reported by other workers. Franks (16) has reported that polyacrylamide gel-entrapped Streptococcus faecalis, which catalyzes
the degradative conversion of arginine to putrescine, was activated by treating it with lysozyme. Chibata et al. (15) activated
Escherichia coli cells with aspartase activity, which was trapped
in polyacrylamide gel, by autolysis.

The pH optimum for CoA synthesis by the gel was between
pH 7-8, but that of the free, dried cells was lower. In the
phosphorylation of pantothenic acid, however, both preparations
showed maximum activity at about pH 6.5. The gel was most
active at 37°C. Even at 45°C, the gel synthesized CoA (75% of the
activity at 37°C), but synthesis by free, dried cells was completely repressed.

Brown (3) has reported that CTP is required to couple P-pantothenic acid with cysteine in a bacterial system. We previously
observed that CTP was superior to ATP as an energy source for the
coupling reaction in the system with free, dried cells. However,
the supplementary effect of CTP on the accumulation of CoA by
these cells was very slight. Conversely, in the immobilized cell
system, remarkable acceleration of CoA synthesis was observed,
especially in the gel used, when CTP was added to the reaction

mixture. This suggests that nucleoside diphosphate kinase is
present in the cells. Probably this enzyme recycles a catalytic
amount of CTP existing in the cells.

A comparison of the immobilized cell method with the other
two methods described is summarized in Table I. As seen in the
table, the fermentative method has the highest productivity based
on three different indexes. The value based on ATP consumption in
the fermentative method clearly demonstrates the recycling of used
ATP. Although the immobilized cell method consumes the most ATP,
its productivity of CoA based on 100 mg of cells is higher than
that in the dried cell method. This, of course, is due to repeated use of the cells.

Isolation of CoA (2,5,6,11,12). The main problem for the
rapid isolation and high recovery of CoA was to separate ATP from
CoA in the cultured broth or the reaction mixture. This was
solved by isolating CoA as the disulfide. The process is outlined
in Fig. 6. Results of actual operations with the three methods
are summarized in Table II. Recoveries of more than 65% were
obtained from the reaction mixtures with free, dried cells or with
immobilized cells.

Table II. Yields and Purities of Isolated CoAs

	Cell method	Fermentative method	Immobilized cell method
CoA content (mg/100 ml)	190	250	213[a]
Column used[b]	charcoal Dowex 1x2	Duolite charcoal Dowex 1x2	charcoal Dowex 1x2
H_2O_2 treatment[c]	done	not done	done
Yield (mg)	150	101	153
Purity (%)	85	83–87	91
Recovery (%)	67	34	65

a) Total yield after 4 times of repeated reactions. For
   detail, see reference 2.
b) See Fig. 6.
c) This process is necessary to oxidize CoA.

Synthesis of the Intermediates of CoA Biosynthesis (2,6,8,12,
16,17,18). CoA and all the intermediates of CoA biosynthesis,
with the exception of P-pantothenoylcysteine, are synthesized
selectively with high yields by the individual reactions involved
in the CoA biosynthetic pathway. The principles of these synthe-
ses are based on the following observations. Only P-pantothenic

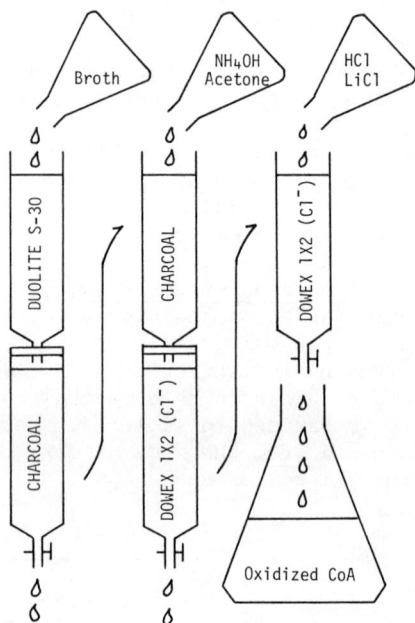

Figure 6. Separation of CoA from the reaction mixture or the culture filtrate. For details, see Ref. 10 and 12.

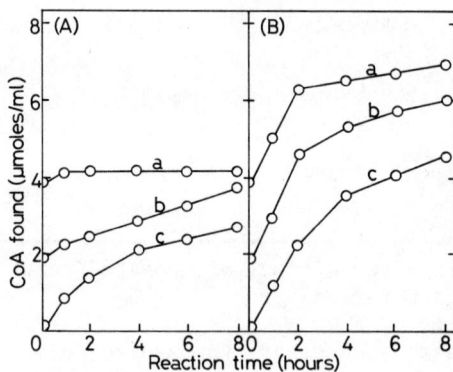

Figure 7. Time course of CoA synthesis in the presence of CoA. (A). Synthesis from pantothenic acid: the reaction was carried out with 4 µmol (a) or 2 µmol (b) CoA. Other conditions were the same as those described in Figure 2(A), except for addition of 2 mg sodium laurylsulfate. The mixture without CoA (c) was used as a control run. (B) Synthesis from P-pantothenic acid: the reaction conditions were the same as those described in (A) except that sodium pantothenate was replaced with P-pantothenic acid. Symbols are the same as those in (A).

Table I. Comparison of Enzymatic Activities in Three Methods

	Cell method	Fermentative method	Immobilized cell method
Productivity			
mg/ml	2–3	3–5	0.5–1.2
mg/100 mg cells	2–3	12–20	3
moles/4 moles ATP	0.7–1.0	2.5–4.3	0.2–0.4
Form of CoA produced	reduced form	oxidized form	reduced form
Source of ATP	ATP	AMP, adenosine or adenine	ATP
Optimal pH for synthesis	6.5	7.6	7.5
Optimal temperature for synthesis (°C)	42	28–30	37
Heat stability[a] (45°C, 3 hours)	10	–	52
Loss of activity during storage	No loss of activity for 2 years (−15°C)[b]	–	No loss of activity for 45 days (0°C)
Surfactant	anionic	cationic or amphoteric	anionic
Activator	–	–	CTP

a) Relative remaining activity (%)
b) Dried cells

acid was synthesized when cysteine was omitted from the reaction
mixture. P-Pantothenic acid most rapidly coupled with cysteine
when CTP was present in place of ATP, but GTP, ITP and UTP were
inactive nucleotides. The reaction with CTP gave P-pantetheine,
and CoA was the main product in the reaction mixture with ATP.
P-Pantetheine was consumed to form CoA only when incubated with
ATP. Other nucleotides lacked the ability to couple with this
substrate. Incubation of pantothenic acid and cysteine with CTP
and ITP, GTP or UTP also gave P-pantetheine. This probably due to
the broad specificity of pantothenate kinase for nucleotides and
the absolute requirement of ATP by dephospho-CoA pyrophosphorylase.
Pantothenate kinase also catalyzed the phosphorylation of pantetheine in the presence of ATP, ITP, GTP or UTP. The incubation of
pantetheine with ITP, GTP or UTP also gave P-pantetheine. Dephospho-CoA was obtained by treating the reaction mixture which had
accumulated CoA with 3'-nucleotidase of Bacillus subtilis.
Although detailed experiments were made using dried cells, these
same principles apply to both the fermentative method and the immobilized cell method. Results are summarized in Table III.

Feedback Inhibition of Pantothenate Kinase by CoA and its
Removal by Continuous Synthesis on an Immobilized Cell Column (2,
6,17,19). The phosphorylation of pantothenic acid is strongly
inhibited by CoA as the end product in the bacterial system as has
also been reported for rat liver pantothenate kinase (20). This
was demonstrated from experiments with dried cells and purified
pantothenate kinase. Probably this feedback inhibition is involved in regulating the intra-cellular CoA level as a general regu-
lation mechanism. However, the presence of this inhibition mechanism is considered undesirable if CoA is to be obtained in a high
yield. This may be why we had difficulty in obtaining a much
higher accumulation of CoA than we did previously. In contrast,
the other enzymes involved in CoA biosynthesis were not sensitive
to CoA even when the concentration of CoA in the reaction medium
was 5 mM. As expected, there was no significant repression of the
amount of CoA synthesized from P-pantothenic acid and cysteine in
4 mM CoA, although synthesis from pantothenic acid and cysteine
was completely repressed under the same conditions (Fig. 7). This
route for the synthesis of CoA is probably one which avoides feedback inhibition. As a remover of feedback inhibition, the immobilized cell method has an advantage over the other two methods in
that synthesis can be produced continuously with a column system.
In this case, the product can be removed continuously from the
reactor. Figure 8 shows the results of the continuous synthesis
of CoA. When fresh reaction mixture was passed through the column
at a flow rate of SV = 0.1-0.2 for 6 days, loss of the initial
activity of the column was about 50%. To remove feedback inhibition completely, we divided the column into two parts. Only the
phosphorylation of pantothenic acid operated at the top of the
column (Fig. 8). The pantothenic acid added was phosphorylated

Table III. Comparison of Three Methods in Productivities of CoA Biosynthetic Intermediates

Product	Substrates	Nucleotides[a]	Enzymes[b]	Productivity (mg/ml)		
				Cell mehtod	Fermentative method	Immobilized cell method
P-Pantothenic acis	Pantothenic acid	ATP (AMP)	1	3-4	4-5	1.5-2.5
P-Pantetheine	P-Pantothenic acid, cysteine	CTP	2,3	3-4		1.8
	Pantothenic acid, cysteine	ITP,CTP (GMP,CMP)	1,2,3	2-3	3-4	0.3
	Pantetheine	ITP (GMP)	1	2-3	4-5	0.9
Dephospho-CoA	Pantothenic acid, cysteine	ATP	1,2,3,4,5	1-2		
CoA	Pantothenic acid	ATP (AMP)	1,2,3,4,5	2-3	3-5	0.5-1.2
	P-Pantothenic acid, cysteine	ATP	2,3,4,5	4.0		0.8-1.7
	Pantetheine	ATP (AMP)	1,4,5	2.5	4.4	1.0
	P-Pantetheine	ATP	4,5	6.7		4.7
	Dephospho-CoA	ATP	5	7.0		5.8

a) Monophosphates in parentheses were used for the fermentative method.
b) 1, pantothenate kinase; 2, phosphopantothenoylcysteine synthetase; 3, phosphopantothenoylcysteine decarboxylase; 4, dephospho-CoA pyrophosphorylase; 5, dephospho-CoA kinase.

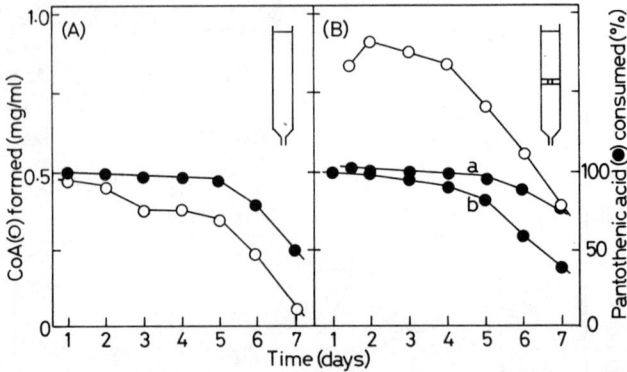

Figure 8. Synthesis of CoA on an immobilized cell column. (A) Single column system: a substrate mixture composed of sodium pantothenate (2.5 μmol/mL), cysteine (10 μmol/mL), ATP (15 μmol/mL), magnesium sulfate (10 μmol/mL), potassium phosphate buffer, pH 6.5 (150 μmol/mL), and sodium laurylsulfate (1 mg/mL) was applied to a column (1 × 20 cm) of gel-entrapped dried cells. The reaction was carried out at 34°C with a flow rate of SV = 0.1–0.2 hr[-1]. (B) Separated column system: a substrate mixture composed of sodium pantothenate (2.5 μmol/mL), ATP (7.5 μmol/mL), magnesium sulfate (10 μmol/mL), potassium phospate buffer, pH 6.5 (150 μmol/mL), and sodium laurylsulfate (1 mg/mL) was applied to the top of the column (1 × 10 cm). Solution (about 20 mL) passed through the column at a flow rate of SV = 0.1–0.2 hr[-1] was collected every day. To the solution (20 mL), 200 μmol cysteine and 150 μmol ATP were added, which was then reacted at the bottom of the column (1 × 10 cm) with a flow rate of SV = 0.1–0.2 hr[-1] to yield CoA. The reaction temperature was 34°C. Consumption of pantothenic acid was checked both at the top (b) and bottom (a) columns.

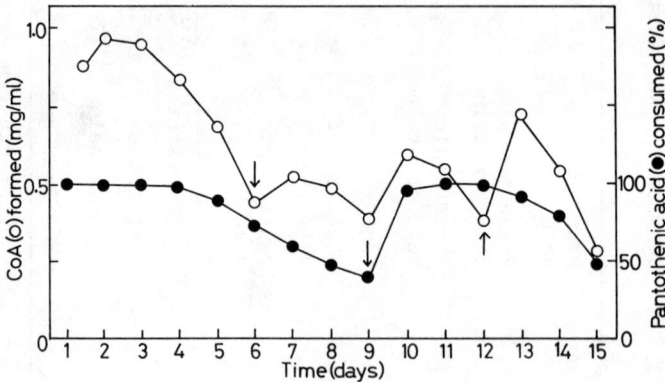

Figure 9. Continuous synthesis of CoA by separated column system. The reaction conditions were the same as those described in Figure 8(B). Column was

almost completely to yield P-pantothenic acid, which then reacted
with cysteine at the bottom of the column to yield CoA. The total
accumulation of CoA after 7 days of operation on this separated
column was about 2-fold higher than that on a single column. A
solution containing more than 500 μg of CoA was obtained continuously for at least 2 weeks when the column was changed 3 times
(Fig. 9).

Conclusion

CoA has been extracted from microorganisms and has been
chemically synthesized. However, these methods are not practical
because of low yields or complexity. The processes described here
are all simple and rapid and require no special equipment. These
processes possess considerable practical advantages over other
previously reported processes in that the product can be obtained
in a high concentration and can be purified in a higher yield with
a compact plant that needs no complex purification procedures.
This not only speeds up isolation and purification but gives a
highly purified product as well.

Abbreviations Used

CoA, coenzyme A; dephospho-CoA, 3'-dephospho-coenzyme A;
P-pantothenic acid, 4'-phosphopantothenic acid; P-pantothenoylcysteine, 4'-phosphopantothenoylcysteine; P-pantetheine, 4'-phos-
phopantetheine; ATP, adenosine 5'-triphosphate; ADP, adenosine 5'-
diphosphate; AMP, 5'-adenylic acid; CTP, cytidine 5'-triphosphate;
CMP, 5'-cytidylic acid; GTP, guanosine 5'-triphosphate; GMP, 5'-
guanylic acid; ITP, inosine 5'-triphosphate; UTP, uridine 5'-tri-
phosphate; PPi, inorganic pyrophosphate.

Aknowledgements

We wish to express our thanks to the late Professor Koichi
Ogata, Kyoto University. Due to his leadership and encouragement
this review has been possible.

Litrature Cited

1. Slowinski, W., and Charm, S.E: Biotechnol. Bioeng., (1973),
   15, 973.
2. Shimizu, S., Morioka, H., Tani, Y., and Ogata, K: J. Ferment.
   Technol., (1975), 53, 77.
3. Brown, G.M., J. Biol. Chem., (1959), 234, 370.
4. Ogata, K., Shimizu, S., and Tani, Y: Agric. Biol. Chem., (1970),
   34, 1757.
5. Ogata, K., Shimizu, S., and Tani, Y: Agric. Biol. Chem., (1972),
   36, 84.
6. Ogata, K., Shimizu, S., and Tani, Y: Proc. 1st Intern. Congr.

IAMS, Tokyo, (1974), 5, 500.
7. Shimizu, S., Tani, Y., and Ogata, K: Agric. Biol. Chem., (1972), 36, 370.
8. Shimizu, S., Miyata, K., Tani, Y., and Ogata, K: Biochim. Biophys. Acta, (1972), 279, 583.
9. Shimizu, S., Miyata, K., Tani, Y., and Ogata, K: Agric. Biol. Chem., (1973), 37, 607.
10. Shimizu, S., Miyata, K., Tani, Y., and Ogata, K: Agric. Biol. Chem., (1973), 37, 615.
11. Ogata, K: "Advances in Applied Microbiology," Vol. 19, ed. by Perlman, D., p.209, Academic Press, New York, 1975.
12. Shimizu, S., Tani, Y., and Ogata, K: "Mehthods in Enzymology," ed. by Colowick, S.P., and Kaplan, N.O., Vol. 62, Academic Press, New York, in press.
13. Tochikura, T., Kuwahara, M., Yagi, S., Okamoto, H., Tominaga, Y., Kano, T., and Ogata: J. Ferment. Technol., (1967), 45, 511.
14. Tanaka, H., Sato, Z., Nakayama, K., and Kinoshita, S: Agric. Biol. Chem., (1968), 32, 721.
15. Chibata, I., Tosa, T., and Sato, T: Appl. Microbiol., (1974), 27, 878.
16. Shimizu, S., Satsuma, S., Kubo, K., Tani, Y., and Ogata, K: Agric. Biol. Chem., (1973), 37, 857.
17. Shimizu, S., Kubo, K., Tani, Y., and Ogata, K: Agric. Biol. Chem., (1973), 37, 2863.
18. Shimizu, S., Kubo, K., Satsuma, S., Tani, Y., and Ogata, K: J. Ferment. Technol., (1974), 52, 114.
19. Shimizu, S., Kubo, K., Morioka, H., Tani, Y., and Ogata, K: Agric. Biol. Chem., (1974), 38, 1015.
20. Abiko, Y., Ashida, S., and Shimizu, M: Biochim. Biophys. Acta, (1972), 268, 364.

RECEIVED March 29, 1979.

Phenol Degradation by *Candida tropicalis* Whole Cells Entrapped in Polymeric Ionic Networks

J. KLEIN and U. HACKEL

Institute of Chemical Technology, Technical University, Braunschweig, F.R.G.

F. WAGNER

Chair of Biochemistry and Biotechnology, Technical University, Braunschweig, F.R.G.

Entrapment in polymeric networks is still the most widely used technique in whole cell immobilization. Such polymeric networks were originally prepared by precipitation -- with-(1) or without (2,3) hardening -- or crosslinking polymerization (4,5). Recently, ionic networks (3,6,7,8) and polycondensation networks (7,8,9,10) have been developed as well.

The ionic network formation procedure was originally developed by Thiele and coworkers (11,12) and our laboratory was the first to adopt and modify this technique to be applicable for whole cell immobilization (3). It was our impression that the contact of the cells with polyelectrolytes and some small electrolytes in aqueous solution only would be most advantageous to maintain high fraction of enzymatic activity and living cells after immobilization.

Phenol degradation by Candida tropicalis has been chosen as a model reaction. As an example of the important reaction class of decomposition of aromatic compounds (13,14,15) it has two characteristic features: (i) the reaction is catalysed by a multienzyme system and (ii) oxygen has to be supplied as a cosubstrate.

Point (i) very likely is related to the problem of immobilizing living (viable) cells while point (ii) directs towards a typical problem in fermentation technology which will be of increasing importance for immobilized cell technology as well. Phenol degradation using isolated enzymes (13,16,17) and free cells (14,18,19,20,21) has been studied in other laboratories, but no work has been done using immobilized cell catalysis.

This contribution on Candida tropicalis immobilization in ionic network polymers is part of a more general study which is reported elsewhere (7,22) involving other immobilization techniques as well.

0-8412-0508-6/79/47-106-101$05.00/0

Experimental Procedures

Microbial Cell Production. A yeast species, classified as Candida tropicalis (Cast.) Berkhout has been isolated and optimized for maximum degradation rate of phenol (in a 1 % solution). Fermentation for biomass production has been performed in a 10 l fermenter, at a stirring rate of 200 rpm, T=30°C and pH=6.5. Aeration rate for the first 6-7 hr (i.e., to the onset of the log phase) was 0.16 N l air/l medium·min, then reset for the remaining 42 hr to 1 N l air/l medium·min. The nutrient medium was composed of 1 g yeast extract, 2 g NH_4NO_3, 2 g $(NH_4)_2SO_4$, 1 g KH_2PO_4, 2 g $K_2HPO_4 \cdot 3H_2O$, 1 g $Na_2HPO_4 \cdot 2H_2O$, 0.2 g $MgSO_4 \cdot 7H_2O$, 0,2 g KCl, 1 ml trace salt solution (Zn^{2+}, Mn^{2+}, Fe^{2+}, Co^{2+}, Cu^{2+}, H_3PO_4, KJ, EDTA) 1 g phenol added to 1000 ml H_2O.
During the fermentation process, seven supplemental additions of 1 g phenol each had to be made, where the timing was given by the stop of yeast cell growth caused by phenol depletion. The cells were separated from the fermentation broth by centrifugation at room temperature, washed twice with 0,9 % NaCl solution and stored at 4 °C. Candida tropicalis cells are spherical   to ellipsoidal particles with a mean diameter of 3 to 5 μm.

Reincubation of Immobilized Cells. Polymer beads with immobilized Candida tropicalis cells were dispersed in a nutrient medium, containing 52 mg K_2HPO_4, 40 mg $MgSO_4 \cdot 7H_2O$, 400 mg $(NH_4)_2SO_2$, 200 mg yeast extract, 200 mg phenol, 40 mg KCl, 0,2 ml trace salt solution and 180 ml H_2O. The fermentation was performed in a 1 l Erlenmeyer flask on a shaking device at 30 °C for a total time of 24 hr. (Eight hours after the onset of the experimantation another 26 mg KH_2PO_4 was added to the mixture.) The polymer beads were seperated and washed extensively to remove cells which might have been grown on the surface. Cell growth within the polymeric network was confirmed on the basis of elemental analysis with regard to nitrogen.

Whole Cell Immobilization. In a screening procedure, different polymers and counterion solutions were studied with regard to their gel formation behaviour. These substances are summarized in Table I, showing that the different types of polymers on the one hand, and $Al_2(SO_4)_3$ solution on the other, are appropriate for cell immobilization. Two of the polymers are available commercially (sodium alginate from Alginate Industries GmbH, Hamburg, and carboxymethylcellulose from Wolff Walsrode AG). The styrenemaleic acid copolymer (copoly(St-MA)) had been prepared in our laboratory following well known procedures (23).

Table I. Survey of Substances Studied
for their Gel Formation Properties

Substance	Classification	Properties + =good,- = poor
Na alginate	Natural polymer	+
Carboxymethylcellulose (CMC)	Seminatural polymer	+
Copoly(styrene-maleic acid) (Copoly(St-MA))	Synthetic polymer	+
Copoly(acrylamide-acrylate)	Synthetic polymer	-
Copoly(trans-stilbene-maleic acid)	Synthetic polymer	-
Copoly(isoprene-maleic acid)	Synthetic polymer	-
Copoly(vinylacetate-maleic acid)	Synthetic polymer	-
Copoly(isobutene-maleic acid)	Synthetic polymer	-
Copoly(vinylmethylether-maleic acid)	Synthetic polymer	-
Copoly(furan-maleic acid)	Synthetic polymer	-
$AlCl_3$	Electrolyte	-
$Al_2(SO_4)_3$	Electrolyte	+
$Al(NO_3)_3$	Electrolyte	(+)
$CaCl_2$	Electrolyte	(+)

The process of polymer bead formation can be divided into five steps: (i) preparation of polyelectrolyte solution in its sodium salt form, (ii) addition of cell mass to (i) and dispersion, (iii) dropping of this suspension through a capillary tube into an Al^{3+} solution, (iv) hardening of beads in the Al^{3+} solution and (v) separation of biocatalyst particles and washing. The capillary device for obtaining a controlled particle diameter has been described elsewhere (7), where a concentric air stream around the capillary tube leads to decreasing particle size with increasing air velocity. While a 0.1 M $Al_2(SO_4)_3$ solution could be applied generally, optimal polymer concentrations had to be determined for each individual polyelectrolyte, and these weight concentrations are indicated in parentheses as follows: Na alginate (c=3%), CMC RIT 5000 (c=1.5%), copoly(St-MA) (c=10%). Typical hardening times in step (iv) were about 30 min.

Toxicity Test. To obtain some information about loss of enzymatic activity from contact of cells with chemicals during the immobilization procedure, cells have been incubated with salt solutions under variation of concentration and incubation time. These values combined with the results on the percentage of

living cells, are given in Table II, and it can be seen that a
survival rate of 80 % under typical preparation conditions seems
to be a good estimate. The data on living cell percentage are
based on agar plate count tests (incubation for 96 h at 30 ^{o}C)
with cells, purified from the salt incubation experiment by centrifugation and washing with pH=7 buffer solution.

Table II Survival of <u>Candida tropicalis</u> cells
during Incubation with Al-Salt Solutions

Salt	Concentration mole/l	Incubation time, min.	Percent living cells
$AlCl_3$	0.5	120	28
$Al_2(SO_4)_3$	1.0	120	12
	0.5	120	13
	0.25	120	31
	0.1	120	71
		96	74
		60	81
		30	83
		20	88
		15	88
		10	90
		5	99

   Kinetic Analysis. A stirred tank slurry reactor was construc-
ted, using a 500 ml polyethylene flask equipped with gas dispers-
ing stirrer and aeration was performed with pressurized air.
Phenol solution (300 ml) was used to suspend 35 ml of catalyst
particles. Phenol conversion was followed by discontinous sampl-
ing and UV analysis at λ=270 nm.
   A fixed-bed reactor device was used otherwise, as shown in
Figure 1. The catalyst bed (1) contained 70 or 140 g of catalyst
beads. Phenol solution was circulated through the bed and an oxygen saturation column (2) from a reservoir (3) of 2 l in volume,
and a circulation rate of 120 l/hr was used to obtain practically
gradient-free conditions in the whole reactor system. This "well
mixed reactor" system could be used in a discontinuous or a continuous fashion, using another reservoir (6) and an overflow channel reservoir into (8) in the latter case. A Zeiss PM QII UV pho-
tometer (4) was connected to the reactor to monitor phenol concentration at λ= 270 nm.
   Oxygen saturation in solution could be established in both
reactors using an oxygen sensing electrode. The substrate solution generally had the following initial composition: 0.282 g
phenol,p.a., 9.0 g NaCl, 0.2 g $MgCl_2 \cdot 6H_2O$, 0.15 g $CaCl_2 \cdot 2H_2O$,
990.37 g H_2O and its pH value was 6.0.

The Phenol Degradation Reaction

Reaction Pathway. The simplest stoichiometric equation for oxydative phenol degradation would, of course, have to be written

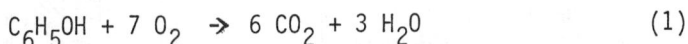

$$C_6H_5OH + 7 O_2 \rightarrow 6 CO_2 + 3 H_2O \qquad (1)$$

However it has been shown by Neujahr and coworkers (24,25) and also by Nei and coworkers (26) that in microbial degradation with Candida tropicalis actually only 3 to 4 moles of oxygen per mole of phenol are consumed. The present knowledge on the phenol degradation pathway can, according to these authors, be formulated as shown in Figure 2. The value of 3.5 for the molar oxygen consumption would correspond to the fact that succinic acid - apart from CO_2 and H_2O - is an end product. The value of 4,however, indicates that succinic acid must be susceptible to further oxidation, as has recently been confirmed in our laboratory (22). In the course of this paper, the value of 4 moles of oxygen per mole of phenol has been used.

Reaction Kinetics for Free Suspended Cells. For the concentration range of phenol covered in this study, change in phenol concentration as a function of time has always been a linear plot, so that at any set of considerations(temperature T, pH, and oxygen concentration (O_2), or yeast concentration) a zero-order rate equation holds:

$$-d(Ph)/dt = \text{Constant} \qquad (2)$$

The specific activity of the Candida tropicalis yeast cells with regard to their phenol degradation capability is therefore a constant, independent of phenol concentration:

$$r_{sp} = (-1/(X)) (d(Ph)/dt) \left[\frac{\text{mole phenol}}{\text{g cells wet weightxh}} \right] (3)$$

The dependence of this specific activity value on temperature and oxygen concentration is shown in Figures 3 and 4, respectively(27). The dotted lines are used to indicate regions where a fast deactivation of cells was observed.

From the data in Figure 3, an apparent activation energy of E_a = 10.2 kcal/mole can be obtained. Figure 4 shows that the reaction becomes oxygen limited at (O_2) <5 mg/l, while otherwise a zero-order dependence holds. In the region of pH = 5.5 to 8, only a very flat maximum of r_{sp} can be observed (27).

Reaction Kinetics for Immobilizes Cells. Evaluation of the catalytic behaviour of the immobilized cells can best be made by comparing the effective values of specific phenol degradation activity with the corresponding values for free suspended cells, thus defining a relative activity (or catalytic efficiency factor)

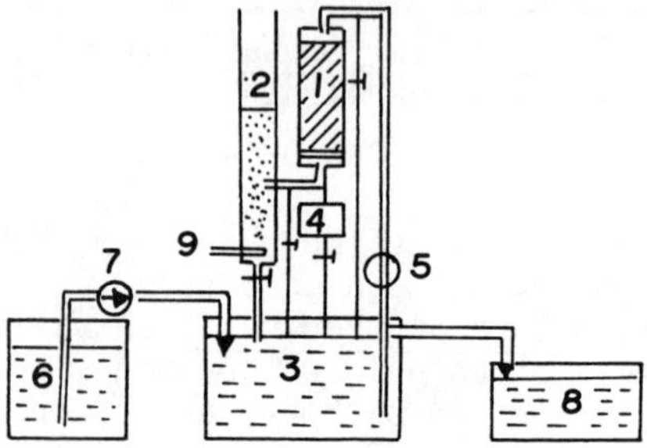

Figure 1.   Circulation reactor for phenol conversion studies: 1, packed bed of catalyst particles; 2, oxygen saturation column; 3, substrate solution reservoir; 4, UV spectrophotometer; 5, circulation pump; 6, substrate solution reservoir for continuous feed; 7, feed pump; 8, product solution reservoir; 9, air distribution device.

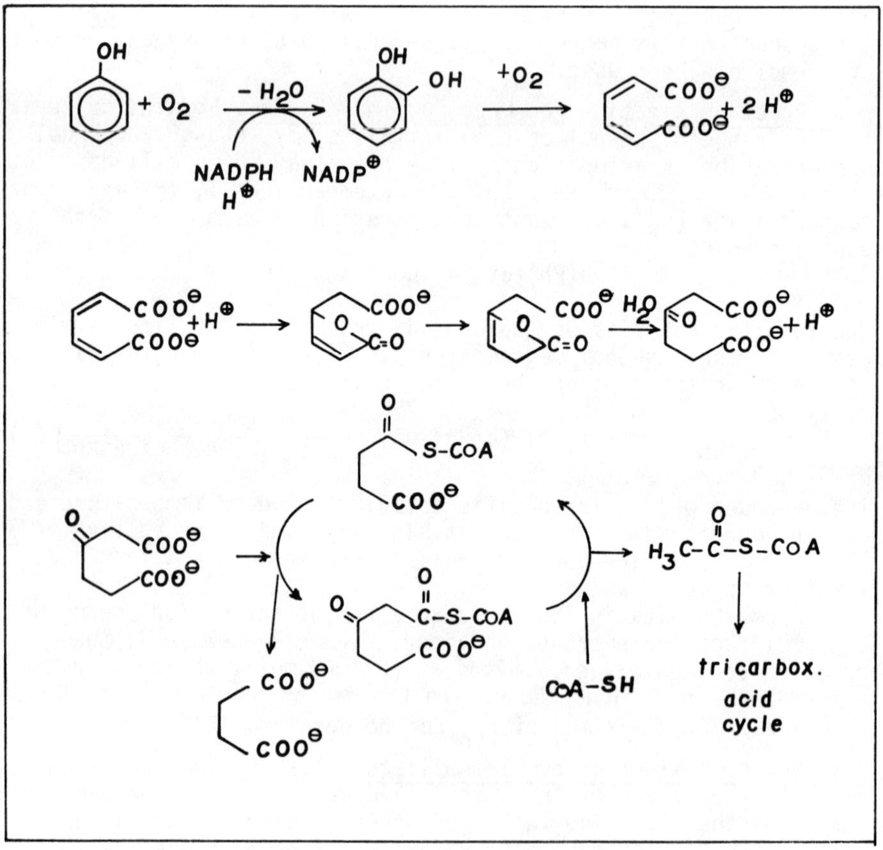

Figure 2.   Proposed reaction pathway for enzymatic, oxidative phenol degradation, according to the literature (21, 24, 25, 26)

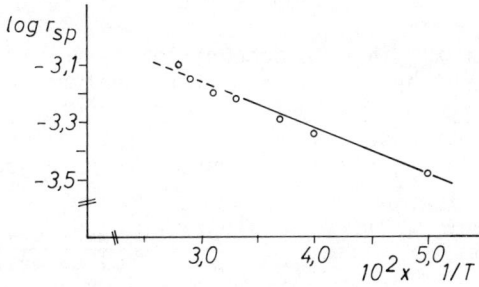

Figure 3. *Specific reaction rate r_{sp} for phenol degradation by suspended cells as a function of temperature*

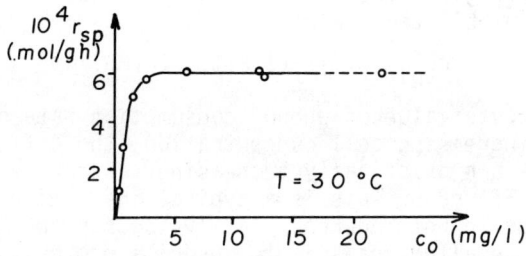

Figure 4. *Specific reaction rate r_{sp} for phenol degradation by free suspended cells as a function of oxygen concentration in solution*

$$\eta = \frac{r_{sp,eff}}{r_{sp}} \tag{4}$$

The calculation of $r_{sp,eff}$ requires axact knowledge of the concentration of catalytically active immobilized cells per unit volume of catalyst particles. The immobilization technique guarantees the complete entrapment of the cells suspended in the polymer solution before crosslinking. If this cell concentration is denoted by

$$X_i = \frac{g \text{ yeast cells wet weight}}{\text{ml polymer solution}}$$

the expression for X_{act} has to account for a) volume changes caused by the crosslinking reaction of the polymeric component and b) for activity loss by toxicity. With respect to a) a volume decrease of 30 % was observed, with respect to b) considering the data in Table II 20 % activity loss seems to be a good estimate, leading to

$$X_{act} = \frac{0.8}{0.7} X_i = 1.14 X_i \left[\frac{g \text{ active yeast cells wet weight}}{\text{ml catalyst}}\right] \tag{5}$$

Typical diagrams for phenol degradation by immobilized cells in discontinuous reaction runs are shown in Figures 5 and 6. The different lines in Fig. 5 refer to different cell concentrations, while in Fig.6 different reaction temperatures are compared. The phenol consumption rate, obtained from the slopes of these lines, divided by X_{act}, gives the effective specific activity of the cells in the immobilized state:

$$r_{sp.eff} = (1/X_{act}) (d(Ph)/dt) \tag{6}$$

While the absolute value of phenol consumption rate obviously increases with increasing cell concentration, the effective specific activities are drastically decreasing, as can be seen from the legend of Figure 6. This is a typical indication for transport limitation in heterogeneous catalysis, a problem which requires a more detailed analysis to describe properly the experimental findings on catalytic efficiency in immobilized cell catalysis (8,22,24).

Catalytical Activity and Transport Limitation
Catalytical Efficiency and Thiele Modulus- As is well known in heterogeneous catalysis, the relevance of transport limitations in such a reaction system can be evaluated if the Thiele modulus is known (28),29). For a set of assumptions (28), the Thiele modulus for spherical particles and a zero-order reaction can be written

$$\phi = (R/3) \sqrt{k_0/2D_{eff}c_0} \tag{7}$$

and ϕ is related to the catalytic efficiency η by

$c \cdot 10^3$ (mol/l)

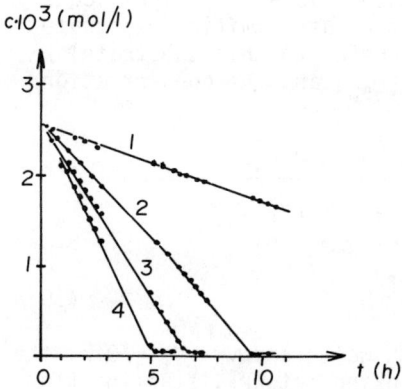

Figure 5. Phenol concentration as a function of time in discontinuous reaction runs with Candida tropicalis *cells entrapped in copoly(styrene–maleic acid) /Al³⁺ networks. R = 1.5 cm; T = 293.15 K; 1, X_i = 0.0035; 2, X_i = 0.02; 3, X_i = 0.06; 4, X_i = 0.10.*

$c \cdot 10^3$ (mol l)

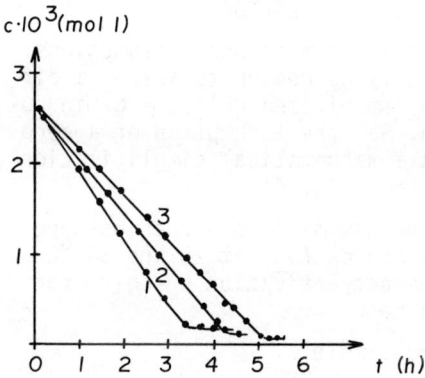

Figure 6. Phenol concentration as a function of time in discontinuous reaction runs with Candida tropicalis *cells entrapped in copoly(styrene–maleic acid) Al³⁺ networks. R = 0.13 cm; X_i = 0.06; 1, T = 303.55 K; 2, T = 299.15 K; 3, T = 292.05 K.*

$$\phi = \frac{1}{3}\left[1 - \frac{2}{3}\eta-(1 - \eta)^{2/3}\right]^{-1/2} \qquad (8)$$

which can be approximated with sufficient accuracy by $\phi = 1/\eta$ for
$\phi > 5$. In eq.(7) R is the radius of the catalyst bead, k_0 is the
rate constant for oxygen consumption, c_0 is the liquid phase oxygen concentration and D_{eff} is the effective oxygen diffusivity in
the polymeric matrix of the catalyst. (At the conditions of these
studies the molar concentration of phenol is always significantly
larger than of oxygen, so oxygen is the rate limiting substrate
and all parameters in eq.(7) have to refer to this substrate) k_0
is related to the specific activity (r_{sp}) and the concentration
(X_{act}) of the catalytic sites by

$$k_0 = \frac{4\, r_{sp}X_{act}}{3600} = k_v X_{act} \qquad \left[\frac{mole\ oxygen}{cm^3 sec}\right] \qquad (9)$$

Values for D_{eff} can be obtained by using eq.(9)

$$D_{eff} = e^{-4v_{p,eff}} D_0 \qquad (10)$$

where D_0 is the oxygen diffusivity in water and $v_{p,eff,i}/0.7$, the
polymer concentration in the crosslinking network, starting from
a solution concentration $v_{p,i}$ and corrected for 30 % volume con-
contraction. We therefore obtain eq. (11)

$$\phi = (R/3)\sqrt{k_v X_{act})/(2D_{eff}c_0)} \qquad (11)$$

for an approximate calculation of ϕ under mentioned assumptions.
It should be made clear, that there is no reason to assume a different kinetic scheme for free and immobilized cells, e.g. in relation to the oxygen concentration. So the assumption of a zero
order reaction is nothing else but a mathematical simplification
of a more complex actual behaviour (see Fig. 4)

    Based on literature sources and assuming an Arrhenius-type
equation to hold, values of D_0 (31) and c_0 (32) at 20 and 30 °C
have been used to estimate the apparent activation energies for
oxygen diffusion and solubility to be

$$E_{a,D} = 3.67\ kcal/mole$$
$$E_{a,S} = -2,81\ kcal/mole$$

Combined with the activation energy for phenol degradation by free
suspended cells of $E_{a,r}=10.2$ kcal/mole, the temperature dependence
of the Thiele modulus can be expressed by

$$E_{a,\phi} = \frac{1}{2}(E_{a,r} - E_{a,D} - E_{a,S}) = 4.68\ kcal/mole$$

which can be used to determine the temperature dependence of η as
well.

Comparison of Calculated and Experimental Results. In an initial series of experiments, different types of polymeric networks were compared in a more qualitative way with regard to their catalytic performance. The packed bed circulation reactor was used to follow the phenol consumption in discontinuous reaction runs. The data tabulated in the Table III show that different polymeric materials can be applied in ionic network entrapment, with some preference to the alginate and copolymer systems, if compared to CMC. It can be seen furthermore, that immobilized cells may well be susceptible to higher bulk oxygen concentration than the free suspended cells, at least for the time periods observed. Comparing the calculated and experimentally determined values of η some discrepancy has to be envisaged, the experimental values being always smaller, however, following generally the trends predicted by the calculated values.

Table III. Comparison of Different Polymer Networks for Entrapment of Candida tropicalis with Respect to Catalytic Efficiency in the Oxidative Degradation of Phenol

Polymer	$v_{p,i}$ g/cm^3	X_i g/cm^3	$10^7 c_0$ mol/cm^3	$10^4 r_{sp}$ mol/gh	η_{exp}	η_{calc}	ϕ_{calc}	R,cm
Alginate	0.03	0.05	11.43	2.1[a]	0.35	0.49	1.78	0.2
	0.03	0.1	11.43	1.6[a]	0.27	0.36	2.52	0.2
	0.03	0.1	11.43	1.3[a]	0.22	0.36	2.52	0.2
CMC, RIT 5000	0.018	0.09	11.43	0.49[a]	0.08	0.39	2.31	0.2
	0.018	0.09	2.35	0.36[a]	0.06	0.19	5.08	0.2
	0.015	0.1	2.35	0.40[a]	0.07	0.18	5.31	0.2
RIT 30	0.075	0.1	11.43	0,76[a]	0.13	0.32	2,86	0.2
Copoly-(St-Ma)	0.10	0.02	11.43	2.8[b]	0.38	1.00	0.57	0.075
	0.15	0.02	11.43	1.5[b]	0.21	0.80	1.14	0.13
	0.03	0.05	11.43	2.4[b]	0.33	0.57	1.48	0.15
	0.036	0.1	11.43	1.2[b]	0.16	0.42	2.13	0.15
	0.03	0.15	11.43	1.1[b]	0.15	0.34	2.57	0.15
	0.10	0.02	2.35	2.0[b]	0.27	0.65	1.26	0.075
	0.03	0.15	2.35	0.45[b]	0.06	0.17	5.66	0.15

a. r_{sp}=6.0x10^{-4} mol/g h.   b. r_{sp}=7.4x10^{-4} mol/g h.

To obtain a more quantitative answer to this discrepancy, another series of experiments was performed, using copoly(St-Ma) as the matrix-forming polymer. The variables in these experiments have then been (i) particle radius R, (ii) cell concentration X,

and (iii) reaction temperature. To avoid problems which might a-
rise from external mass transfer limitations, the suspended particle reactor type has been used and oxygen was supplied by saturation with air under atmospheric pressure. The data listed in
Table IV show an improved correspondence between η_{exp} and η_{calc} if
compared to Table III; however, the general trend for deviation
remains.

Keeping in mind the set of simplifying assumptions on which
the calculatory process is based, these deviations between model
and experiments do not seem to be too dramatic. Specifically,(i)
the homogeneity of network, (ii) the zero-order reaction kinetics
and (iii) the procedure to estimate D_{eff} may be mentioned in this
context, but no attempt was made in this study to develop a more
refined mathematical model.

Plotting all the data from Table IV in such a way to compare
the Thiele moduli from model calculation ϕ_{calc} against the experimental values ϕ_{exp} in a double logarithmic diagram (see Fig.7),
a two-parameter correlation equation

$$\phi_{exp} = 1.44 \, \phi_{calc}^{1.24} \qquad (12)$$

can be derived, which may serve as a basis for process design calculations for this specific system.

Catalytic Stability and Cell Viability

Continuous Reaction Run. A sufficient stability of catalytic
activity with respect to time on stream is an obvious prerequisite
for any practical application. Using the packed bed circulation
reactor in its continuous mode, a catalytic activity profile, as
shown in Figure 8, was obtained. Candida tropicalis cells (20 g)
had been immobilized using a $v_{p,i}$=0.075 g/ml CMC RIT 30 solution,
leading to a value of X_i=0.1 g/ml. A $3x10^{-3}$ molar solution of
phenol was continuously fed to the system, while oxygen saturation
was obtained with pure oxygen gas.

The lines 1 to 6 indicate a procedure of washing and repacking of the catalytic bed, since settling and compression of the
bed obviously gave rise to increasing pressure drop and decreasing substrate (especially oxygen) supply of the catalyst beads.

Considering the fact, that the procedure of washing and repacking could be completed in about 1 h from the overall average
decay of catalytic activity indicated by the broken line a catalytic half-life of $t_{1/2}$ = 19 days could be determined.

Catalyst Reincubation. It is one of the unique features of
using whole cells as immobilized catalysts that the catalyst can
be activated by reincubation as long as viable cells are present
in the entrapped state. Reincubation of the catalyst beads with
the appropriate nutrients facilitates the growth of new cells inside the polymer bead. This technique can be used either to improve the catalytic activity directly following immobilization or

Table IV. Catalytic Efficiency for Phenol Degradation of Candida tropicalis Cells Entrapped in Copoly(styrene-maleic acid)/Al^{3+} Networks for Various Particle Radius R, Initial Cell Concentration X_i and Reaction Temperature T

	T $^\circ$K	$r_{sp,eff}$	η_{exp}	η_{calc} [a]	R cm	X_i g/cm^3
1	303.35	3.1	0.42	0.63	0.09	0.015
	296.75	2.5	0.49	0.72	0.09	0.015
	293.15	2.1	0.51	0.77	0.09	0.015
2	303.15	1.5	0.21	0.35	0.09	0.06
	299.85	1.5	0.25	0.38	0.09	0.06
	296.85	1.0	0.20	0.41	0.09	0.06
	293.15	1.1	0.27	0.44	0.09	0.06
3	302.95	0.45	0.06	0.20	0.09	0.2
	293.15	0.39	0.10	0.26	0.09	0.2
4	303.55	1.1	0.15	0.25	0.13	0.06
	299.15	0.84	0.14	0.27	0.13	0.06
	292.05	0.59	0.15	0.33	0.13	0.06
	296.75	0.85	0.17	0.25	0.15	0.06
5	308.15	1.0	0.11	0.22	0.13	0.06
	298.75	0.87	0.15	0.28	0.13	0.06
	290.35	0.52	0.15	0.34	0.13	0.06
	303.15	0.71	0.10	0.22	0.15	0.06
6	301.95	4.5	0.65	0.67	0.15	0.005
	301.95	2.6	0.37	0.42	0.15	0.015
7	303.15	6.3	0.84	0.74	0.15	0.0035
	303.15	2.4	0.32	0.36	0.15	0.02
	303.15	1.0	0.14	0.22	0.15	0.06
	303.15	0.6	0.08	0.17	0.15	0.1
8	293.15	2.0	0.49	0.88	0.15	0.0035
	293.15	1.2	0.29	0.46	0.15	0.02
	293.15	0.6	0.13	0.28	0.15	0.06
	293.15	0.4	0.09	0.22	0.15	0.1
9	303.15	0.65	0.09	0.17	0.15	0.1
10	295.95	2.8	0.57	0.84	0.15	0.0035
	298.35	0.56	0.10	0.19	0.15	0.1

a from eq. (8) and (11)

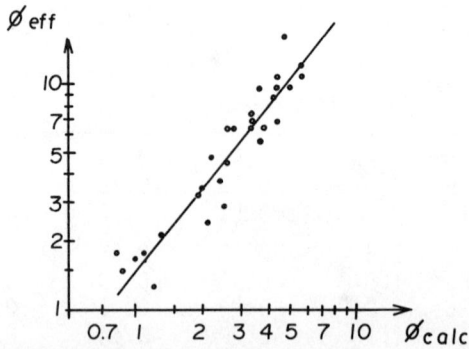

Figure 7. Comparison of values of the Thiele modulus obtained from model calculations (Φ_{calc}) and from experiment (Φ_{exp}) for phenol degradation by Candida tropicalis *immobilized by entrapment in copoly(styrene–maleic acid)/Al^{3+} networks*

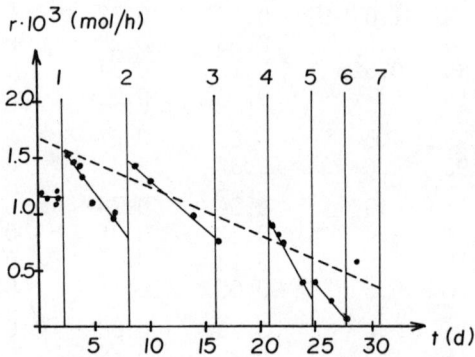

Figure 8. Phenol degradation activity of Candida tropicalis *cells immobilized in CMC/Al^{3+} gels in a continuous reaction run as a function of time of operation: $v_{p,i} = 0.075$ (g/mL); 20 g yeast cells; $X_i = 0.1$; $c_0 = 1.1 \times 10^{-6}$ [mol O_2/mL]; $c_{Ph} = 3 \times 10^{-3}$ (mol/L); $V = 0.6$–0.7 (L/hr); $T = 303$ K. 1–6, Washing and repacking of catalyst bed; 7, end of test.*

to reinstall the initial activity of a catalyst preparation after some deactivation under process conditions. In general, therefore, not only the specific activity but also the lifetime $t_{1/2}$ of a catalyst preparation can be significantly improved.

Table V provides the characteristic data of some reincubation experiments which prove that the concept of reincubation of viable cells is a realistic one.

Table V. Experimental Results for Composition and Catalytic Performance of Candida tropicalis Entrapped in Copoly(styrene-maleic acid)/Al^{3+} Networks Before and After Reincubation with Nutrients

X_i	R, cm	% of nitrogen				$10^4 r_{sp,eff}$	
		Before reincubation	After reincubation	X(after reincu)/X_i	Increase in absolute phenol consumption (factor)	Before reincubation	After reincubation
0.015	0.09	0.31	3.45	11.1	1.48	2.55	0.34
0.06	0.09	1.21	3.56	2.9	2.03	1.13	0.78
0.06	0.15	1.49	3.19	2.1	1.58	1.23	0.91
0.06	0.15	1.39	3.01	2.2	2.45	0.61	0.69
0.2	0.09	4.57	6.94	1.5	1.24	0.27	0.22

Different catalyst preparations, obtained by Candida tropicalis entrapment in copoly(St-MA) networks, as studied above (see Table IV) were used in this experiment, so that the content (%) of nitrogen (on dry weight basis) could be used to calculate the increase of cell concentration following reincubation.

Depending on the initial cell concentration, an increase of cell concentration up to a factor of 11.1 could be obtained. Increase in cell concentration X_{act} generally enhances phenol degradation, however, the increase of X_{act} leads to an increase of the Thiele modulus ϕ, therefore, decreasing the catalytic efficiency η. So the maximum increase of absolute phenol consumption rate of 2.5 is not at all coupled to the maximum value cell concentration increase (see Table V).

A semiquantitative calculation of the catalytic efficiency factor is the only way to derive rational criteria for the optimal strategy of a reincubation step.

Acknowledgement

The financial support of this work by the Federal Department of Research and Technology (B.M.F.T.) of the Federal Republic of Germany under grant no. BCT 65 is gratefully acknowledged.

Abstract

Entrapment of Candida tropicalis whole cells in ionic polymeric networks has been achieved using different types of polyelectrolyte polymers (alginates, carbomethylcellulose, synthetic maleic acid copolymers) and multivalent counterions (e.g.,Al^{3+}). Oxidative degradation of phenol served as a model reaction for a multienzyme catalytic process.

Quantitative analysis of the overall performance of the immobilized cells has been made, thereby distinguishing the effects of (i) deactivation of cells during immobilization, (ii) diffusional limitation with regard to oxygen supply and (iii) catalytic cell stability. Analysis of (ii) is based on the Thiele modulus approach, well known in heterogeneous catalysis. Considering the inherent difficulties in obtaining the correct system parameters, calculated and experimental values of catalysis efficiency agree reasonably well. The special feature of (iii) is the immobilization of viable cells and data are presented which give clear evidence of the growth of new cells within the polymer network, if immobilized cell particles are reincubated with required nutrients.

Literature Cited

1. Vieth, W.R., Wang, S.S. and Saini,R.,Biotechnol. Bioeng. (1973) 15, 565
2. Dinelli, D., Process Biochem. (1972) 7: 9-12
3. Hackel,U.,Klein,J.,Megnet,R. and Wagner,F.,Eur. J.Appl.Microbiol. (1975) 1, 291
4. Mosbach,K. and Mosbach,R.,Acta Chem.Scand.(1966) 20, 2807
5. Yamamoto,K. Sato,T., Tosa,T. and Chibata,I.,Biotechnol.Bioeng.(1974) 16,1589
6. Kierstan, M. and Bucke, C., Biotechnol. Bioeng.(1977)19,387
7. Klein, J. Hackel, U. Schara,P. and Eng,H., Angew. Makromol. Chem. (in press)
8. Klein, J., Hackel,U., Schara,P., Washausen,P., Wagner,F.and Martin,C.K.A., Enzyme Engineering,Vol.4, Plenum Publ.Corp. 1978, p. 339
9. Klein, J., Washausen,P., Eng,H. and Wagner,F., 1st European Congress on Biotechnology, Interlaken 1978 (discussion Paper 213)
10. Klein, J. and Wagner,F., 1 st European Congress on Biotechnology, Interlaken 1978 (survey paper "Immobilized Whole Cells"
11. Thiele,H., Kolloid. Z., (1954),136, 80
12. Thiele,H. and Awad,A. J. Biomed. Mater. Res. (1969)3,431
13. Varga,J.M. and Neujahr,H.Y., Eur. J. Biochem. (1970)12,427
14. Shimizu,T., Uno,T., Dan,Y., Nei,N. and Ichikawa,K., J. Ferment. Technol. (1973) 51,809
15. Somerville,H.J., Mason,J.R. and Ruffel,R.N., Eur. J. Appl. Microbiol.(1977) 4,75
16. Neujahr,H.Y. and Gaal,A., Eur. J. Biochem.(1973) 35,386
17. Neujahr,H.Y., Proc. 5th Internat'l.Ferment. Symposium, Berlin, 1976, p.428
18. Gibson,D.T., Koch,J.R. and Kallio,R.E., Biochemistry(1968)7, 2653
19. DerYang,R. and Humphrey,A.E., Biotechnol.Bioeng.(1975)17, 1211
20. Nei,N., J. Ferment. Technol.(1972) 50,536
21. Stanier,R.Y. and Ornston,L.N., Adv. Microbial Physiol.(1973) 9,89
22. Schara,P., Ph.D. Thesis, Technical University, Braunschweig, F.R.G., 1977
23. Alfrey,T.Jr., Bohrer,J.J. and Mark,H.,"Copolymerization", Interscience, New York, 1952
24. Neujahr,H.Y., Proc. IV IFS: Ferment. Technol. Today (1972), 321
25. Neujahr,H.Y.,Lindsjo,S. and Varga,J.M., Antonie van Leeuwenhoek (1974) 40,209
26. Nei,N. Enatsu,T. and Terui,G., J. Ferment. Technol.(1973)51, 803

27. Washausen,P., Dipl. Thesis, Technical University, Braun-
     schweig, F.R.G., 1976
28. Weekman,V.W. Jr. and Gorring,R.L., J. Catalysis (1965)4,260
29. Marsh,D.R., Lee,Y.Y. and Tsao,G.T., Biotechnol. Bioeng.
     (1973) 15,483
30. White,M.L. and Dorion,G.H., J. Polym. Sci. (1961) 55,731
31. Landolt-Börnstein, "Zahlenwerte und Funktionen", 6. Aufl.,
     Vol. II 5a, p. 611, Springer-Verlag, Berlin, 1969
32. Landolt-Börnstein, "Zahlenwerte und Funktionen", 6. Aufl.,
     Vol. II 2b, p. 1-20, Springer-Verlag, Berlin, 1965

RECEIVED March 29, 1979.

Facile Methods for the Immobilization of Microbial Cells without Disruption of their Life Processes

J. F. KENNEDY

Research Laboratory for the Chemistry of Bioactive Carbohydrates and Proteins, Department of Chemistry, University of Birmingham, Birmingham B15 2TT, England

The immobilisation of enzymes by attachment to water-insoluble material has now received considerable attention for some time and possible applications have been pursued extensively (1, 2). A logical extension of this approach, especially where multi-stage enzymic reactions are being considered, is the immobilisation of microorganisms, which are often the source of many enzyme preparations. The advantages of such an approach are immediately obvious. The tedious and time consuming procedures for enzyme extraction and purification are instantly eliminated, cofactors and coenzymes are readily at hand, the cellular enzymes are often organised into the requisite metabolic pathways and problems associated with enzyme instability may also be avoided. Furthermore, the use of immobilised cells would avoid the problem in industrial processes of separating the product from the enzyme.

Hydrous Metal Oxides as Supports

Justification. Investigation of a number of gelatinous hydrous metal oxides (frequently called hydroxides, although their full structures are uncertain) has established (3) that hydrous titanium (IV), zirconium (IV), iron (III), vanadium (III) and tin (II) oxides at least are capable of forming with enzymes insoluble complexes which are enzymically active. From the practical viewpoint hydrous titanium (IV) and zirconium (IV) oxides proved the most satisfactory. Comparatively high retentions of enzyme specific activity may be achieved (3, 4, 5). Such hydrous metal oxide materials have also proved to be suitable for the immobilisation of amino acids and peptides (3), antibiotics with retention of antimicrobial activity (6), polysaccharides (7), etc.

0-8412-0508-6/79/47-106-119$05.00/0
© 1979 American Chemical Society

Hydrous titanium (IV) and zirconium (IV) oxides are insoluble over the normal physiological pH range, and since when acting as enzyme immobilisation matrices they give good retention of enzymic activity, they thus seem to have little or no effect on the function of biologically active molecules. If enzymic activity, which is extremely sensitive to conformational changes in the enzyme molecule, is not seriously affected by immobilisation, then there seems to be little reason why cell walls should be disrupted or destroyed by this process and the cells themselves, therefore, have a good chance of remaining viable.

Mechanism for Formation of the Cell-Support Bond. The immobilisation process for the hydrous metal oxides is envisaged as involving the replacement of hydroxyl groups on the surface of the metal hydroxide by suitable ligands from enzyme or cell, resulting in the formation of partial covalent bonds. In the case of enzymes, such ligands could be the side-chain hydroxyls of L-serine or L-threonine, the carboxyls of L-glutamic acid or L-aspartic acid and the ε-amino group of L-lysine residues, (illustrated in Figure 1, using hydrous zirconium (IV) oxide as example), oxygen-containing ligands being preferred to those containing nitrogen. In the case of cells, the structural complexity of the cell wall ensures the availability of a great diversity of suitable ligands from both protein, and also from carbohydrate moieties (illustrated in Figure 2, using hydrous titanium (IV) oxide as example).

Experimental Method for Formation of the Support. Samples (each of 1.3 mmol) of the metal hydroxides for use in cell immobilisation were prepared from solutions of their tetrachlorides (titanium (IV) chloride 15% w/v in 15% w/v hydrochloric acid (BDH, Poole, England) and zirconium (IV) chloride (BDH) 0.65 M in 1.0 M hydrochloric acid) by the slow addition of 2.0 M ammonium hydroxide to neutrality (pH 7.0). The samples were washed with saline solution (0.9% w/v, 3 x 5.0 ml) to remove ammonium ions and then used for cell immobilisation studies as below.

Experimental Method for Immobilisation of Cells. The procedure for cell immobilisation is very simple and is illustrated by the following example. A suspension (in 10 ml 0.9 w/v saline) of Escherichia coli cells (A_{600} 0.216) was mixed with a sample of the metal hydroxide as prepared as above (pH 5-7) and agitated gently for 5 min at room temperature. The mixture was then allowed to stand at room temperature and the suspension settled out, leaving a clear supernatant

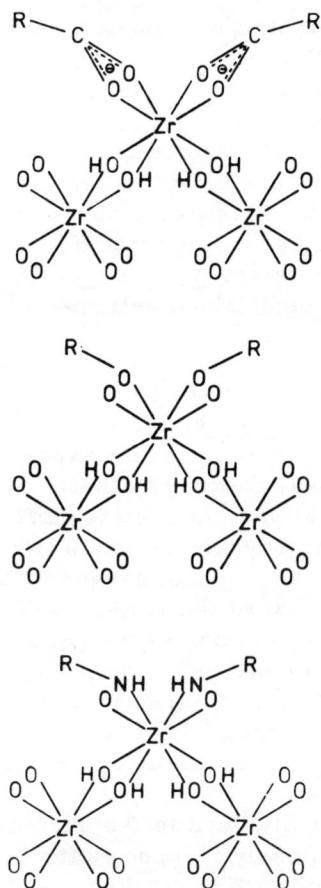

Figure 1. *Projected structures of the chelates/complexes of hydrous zirconium (IV) oxide with carboxyl, hydroxyl, and amino groups, respectively*

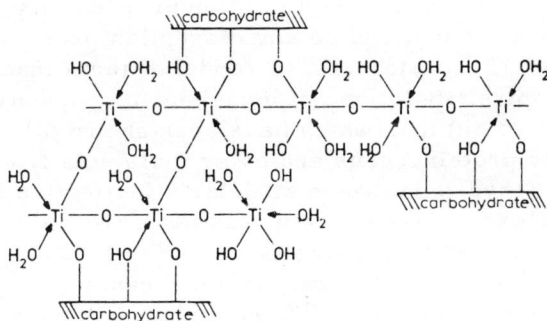

Figure 2. *Projected structure of the chelate/complex of hydrous titanium (IV) oxide with macromolecular carbohydrate*

$(A_{600nm}^{1.0cm} 0.022)$ which was practically devoid of microorganisms
(shown by microscopy). The immobilised cell preparation was
consolidated by centrifugation at low speed and removed from
the supernatant for further examination.

It was also found that cells could be immobilised at lower
pH, hydrous titanium (IV) oxide being more effective for this
purpose, in the range studied (pH 2-5). This phenomenon is
useful since not all microorganisms exist in a neutral pH
environment (for example, Lactobacillus and Acetobacter) and
enabled us to produce a small scale immobilised cell reactor,
as is now described.

Demonstration of the Life etc. Characteristics of

Immobilised Cells. Saccharomyces cerevisiae and E. coli cells
were immobilised in this manner and the preparations were
examined for continued viability by measurement of their
oxygen uptake (at 25°C in aerated 0.2 M sodium acetate buffer
pH 5.0 or 0.9% w/v saline) by use of an oxygen electrode (Pye
Unicam Ltd., Cambridge, England). The rate of oxygen uptake
of the immobilised cells was ~30% of that of the same number of
free cells. This result showed that respiration of the cells
could continue when the cells were immobilised. The reduced
rate of oxygen uptake is probably caused by the restriction by
the hydrous metal oxide of access of aerated buffer to the cells
and a decrease of the area of cell surface available for oxygen
transfer.

To show that the cells were firmly attached to the surface of
the hydrous metal oxide and not just loosely trapped in the
gelatinous matrix, a number of arguments are invoked. Free
cells of E. coli are comparatively small and cannot be
centrifuged down to any significant extent in the centrifugation
conditions used for collecting the immobilised cells; they are
not particularly robust and so any disruption process which had
occurred would, therefore, have rendered them inactive and so
unable to respire. Solutions of bicarbonate, phosphate,
fluoride (and so on) ions which have been shown (3) to remove
loosely bound proteinaceous and other materials from
zirconium (IV) hydroxide were singularly ineffective in the
attempted release of the immobilised cells from the matrix.

To show further that the cells were firmly attached to the
hydrous metal oxide, a different microorganism, Serratia
marcescens was employed. When incubated in nutrient medium
at 25°C, this organism produces a distinctive red colouration,
which enables the immobilised cells to be distinguished readily

from those of any other contaminating microorganisms and also
obviates the need for sterile experimental conditions.

On adding cultures of S. marcescens to hydrous zirconium
(IV) and titanium (IV) oxides, the red coloration (that is, the
cells) became associated with the insoluble matrix, the supernatant and subsequent washings (with saline solution) being
almost cell free, thus demonstrating the strength of the cell-
metal hydroxide interaction. When samples of these immobilised cells were added as a small inoculum to fresh culture
medium, growth was observed (as detected by the large increase
in number of red cells). Since at no time throughout extended
studies of the growth of S. marcescens, in various conditions of
humidity, oxygen tension, temperature and medium composition
and ultraviolet irradiation-induced mutation, was release of
colour from the cells observed, this was taken as evidence that
the cells had suffered no deleterious effects on immobilisation.
It is also clear that the immobilised cells can retain their
activity for a matter of weeks/months at least.

Applications of Living Immobilised Cells to Fermentations

   Introduction and Methods of Improving Fermentation Rates.
The manufacture of vinegar, which is based on the actions of
Acetobacter species, is an industry which operates with low
profit margins and this results in there being little investment
in research or modern equipment. This enforced conservatism
means that the inefficient 'Quick' process is still widely used by
many small manufacturers because more efficient, continuous
or semi-continuous method e.g. the Fring's process (8), would
be prohibitively expensive for them to install. What is required,
therefore, is a system which is efficient, simple to operate and
inexpensive to install and run. The tower fermenter fulfils the
first two criteria - it is at least as efficient as the Fring's
stirred tank reactor and simpler in construction but the initial
cost is still comparatively high. To reduce the initial capital
expenditure would require a reduction in the size of the fermenter with a concomitant increase in its volumetric efficiency
(V.E., effluent volume per day/fermenter volume), which is
low for the acetification process (0.5 commercially, 0.8 under
laboratory conditions). High volumetric efficiencies have been
obtained with other organisms, growing in tower fermenters (9,
10) e.g. some yeasts and moulds can be grown at volumetric
efficiencies of 5-10. The reason for this has been attributed to
the flocculent nature of these organisms, which results in a
high concentration of organism in the fermenter and therefore

allows high flow rates to be achieved. Unfortunately bacteria are non-flocculent and this places a constraint upon the speed of operation of the acetification fermenter for, above a certain flow rate, bacteria are 'washed out' of the fermenter at a faster rate than their growth rate and thus the bacterial concentration and hence the amount of acetic acid produced fall. Thus, if the fermenter is to operate at high flow rates, some means must be found of maintaining an increased bacterial concentration in the fermenter. Since, under a given set of conditions, the growth rate of the organism has a maximum value which cannot be exceeded, the only alternative solution would seem to involve treating the bacteria in some way such that they have a greater tendency to remain in the fermenter, whilst at the same time retaining their ability to convert ethanol into acetic acid.

Answers to the Problem of the need for Increased Bacterial Concentration in a Fermenter. One obvious approach to solving this problem would be to induce the bacteria to flocculate thus, hopefully, giving them a greater resistance to washout in an analogous manner to flocculated yeast cells. However, whilst flocculent strains of yeasts occur naturally, even aggregating strains of Acetobacter do not aggregate to any marked extent and thus some artificial means is required to bring this about. Immobilisation of the cells without harm or damage, according to the above method, was predicted to achieve this desired increase in cell concentration. Bacterial cell walls are notable for their chemical complexity and diversity and the cell walls of Acetobacter (Gram-negative) could be expected to contain mureins (peptidoglycans) as the innermost layer of the wall, with proteins, lipoproteins and lipopolysaccharides in the outer layers. Outside the cell wall there may be a capsule or slime layer of simple composition consisting of either a single polysaccharide or a polypeptide containing a single amino acid. Thus on the basis of the foregoing there seemed ample opportunity for chelation of the cells to hydrous titanium (IV) and zirconium (IV) oxides.

The use of hydrous titanium (IV) oxide rather than hydrous zirconium (IV) oxide seemed to be necessary because of the conditions prevailing in the fermentation. The acetification of ethanol takes place at about pH 3, in which region hydrous zirconium (IV) oxide is not completely precipitated i.e. there are still many zirconium (IV) ions in solution. Hydrous titanium (IV) oxide, by comparison, is fully precipitated at this pH and results obtained on the insolubilisation of starch by hydrous titanium (IV) oxide indicate that the complexing effect begins to be significant in the region of pH 3 (3). This is a very useful property since it indicates that there will be an equilibrium existing between immobilised and free bacteria. One result of this is that when the immobilised cells die, they will eventually be released from the hydrous oxide surface and be replaced by living cells. Thus, the hydrous oxide will not become rapidly exhausted by being covered with dead cells. Another consequence of this is that if the immobilised cells are incapable of reproduction then there will be sufficient free cells to grow normally and replace those bacteria dying or being washed out of the fermenter in the effluent.

Practical Demonstrations of the Use of Living Immobilised Cells for Increasing Fermenter Efficiencies. The tower fermenter system used is described in Figure 3. Wort was used as the ethanol source, and an inoculum of an aggregating strain of Acetobacter species was prepared and added to the tower fermenter ($2\frac{1}{2}$ litres capacity). When the level of acetic acid in the fermenter had reached about 3% w/v, the medium delivery pump was started and the flow rate adjusted to a level that gave almost complete conversion of the ethanol available into acetic acid. Undue haste in increasing the flow rate and also serious decrease or stoppage of the air flow caused the expected fall in conversion efficiency. Adjustment of the flow and aeration rate showed that a maximum V. E. of 0.82 could be attained (see Table I).

Table I.

Average Rates of Production of Acetic Acid for Maximum Efficiencies for Aggregating Acetobacter Strains

Average taken over days	Highest efficiency (V. E. x %)	g acetic acid produced per day	Comment
21-29	0.82 x 86 (day 23)	87 ± 7	No hydrous titanium (IV) oxide
81-88	1.64 x 99 (day 88)	263 ± 13	Hydrous titanium (IV) oxide added, $\equiv$ average 0.75 g $TiCl_4$ daily

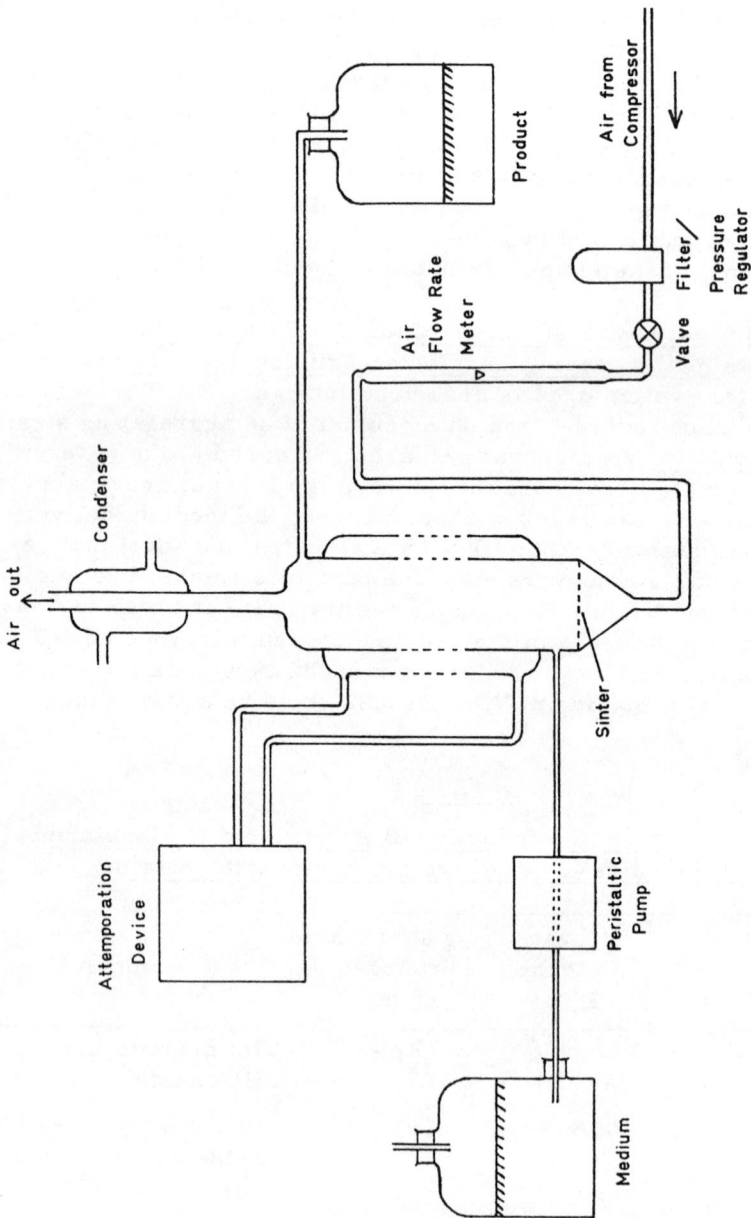

Figure 3. The tower fermentation system

At this point the addition of hydrous titanium (IV) oxide
(prepared as before) began, and over a period the volumetric
efficiency was increased (Table I) the aeration rate being only
slightly increased. At one stage the amount of acetic acid
produced began to fall and this was attributed to the aeration
rate becoming limiting. Consequently, the aeration rate was
increased, to test this hypothesis. That the percentage conversion rose once more, even though the medium flow rate had
increased as well, was taken to be sufficient proof of this.
Ultimately a V. E. of 1.64 was achieved (Table I). Thus
throughout the run it was possible to increase the flow rate and
V. E. whilst maintaining the conversion of ethanol to acetic acid
at between 90 and 100% for the majority of the time.

   The hydrous titanium (IV) oxide had an immediately
noticeable effect on the appearance of the bacteria in the
fermenter, which was to cause a colour change in the organisms
from purple to brown. This did not appear to affect the
performance of the fermenter in any way, however, in
addition, another, more gradual change was observed over a
period of about ten days after the commencement of hydrous
titanium (IV) oxide addition. Before this point, there had been
very little aggregation of the bacteria, although a few small
clumps had formed. After the treatment with hydrous
titanium (IV) oxide many more aggregates formed, dispersed
throughout the fermenter. These were distinguishable in the
close-up photographs of the tower as diffuse spherical
particles, surrounded by air bubbles. A sample removed from
the fermenter and suspended in a glycerol/water mixture,
showed these particles to be spiky pellets, about 2 mm in
diameter. A single particle fixed by heating and stained with
Methylene Blue revealed the presence of particles of titanium
(IV) oxide embedded in the aggregate, visible as dense, dark
spots. This is not a completely true representation of the
situation, since the sample has to be dried before staining
thereby causing the hydroxide particles to shrink and to be
partially converted to oxide. However, it is certain from this
that hydrous titanium (IV) oxide is present in the aggregated
particles, and therefore that the hydrous metal oxide does have
a useful aggregating effect on the bacteria as predicted. Under
higher magnification of the undried sample it could be seen
that many bacteria were not associated with the aggregated
particle (free bacteria), confirming the predicted equilibrium
situation.

   A variety of conditions of fermentations with <u>Acetobacter</u>

species using immobilised cells have now been investigated in various sizes of fermenter. In the case of use of a non-aggregating strain of Acetobacter species ($2\frac{1}{2}$ litre fermenter), whereas addition of hydrous titanium (IV) oxide resulted in an increased V. E. (Table II), this was not so marked as that achieved using the aggregating strain. However, a higher V. E. (Table II) was achieved using hydrous titanium (IV) oxide-cellulose chelate as the support material for cell immobilisation. Hydrous titanium (IV) oxide-cellulose chelate was prepared by mixing equal weights (1.2 g) of chromatographic grade cellulose powder (Whatman CF11) and titanium (IV) chloride solution (as above) and stirring for 2 hours. The mixture was then dried at 45°, ground to a powder, washed with distilled water until the washings were neutral, and then added to the fermenter as an aqueous suspension. The effect of the chelate was not produced by a mere mixture of cellulose and hydrous titanium (IV) oxide. The added success with the chelate is therefore attributable to the altered mode of presentation of the immobilising titanium species.

Conclusions. It may be concluded that the addition of hydrous titanium (IV) oxide to a bacterial fermentation will cause aggregation of the bacteria and thus give a higher bacterial concentration in the fermenter. This apparently allows both a higher aeration rate and a higher dilution rate, giving dilution rates (V. E. ~1.6) at least twice as high as those previously obtained with air aeration (V. E. ~0.8) and comparable with those obtained when the fermenter was aerated with pure oxygen (V. E. 1.8).

More generally, it may be concluded that hydrous transition metal oxides are effective matrices for enzyme, etc. immobilisation. Their advantages include low cost, convenience of preparation (which may be conducted in any location without specialised facilities), the absence of any need for pre-preparation, ability to couple enzyme at neutral pH, high retention of enzyme specific activity of the enzyme on immobilisation, and the ability of modification to exert micro-environmental effects on and thereby alter the characteristics of the immobilised enzyme.

If this simple means of cell immobilisation were applied to other microorganisms it could well result in further immobilised cell reactors of this sort, for the selective production of commercially important biochemical and pharmaceutical compounds. Magnetic forms of the hydrous

Table II

Average Rates of Production of Acetic Acid for Maximum Efficiencies for Non-Aggregating Acetobacter Strain

Average taken over days	Highest efficiency (V. E. x %)	g acetic acid produced per day	Comment
4-14	1.03 x 99 (day 9)	178	No hydrous titanium (IV) oxide
15-17	1.00 x 95 (day 17)	185	Hydrous titanium (IV) oxide added, 0.6 g $TiCl_4$ daily
18-24	0.98 x 97 (day 19)	191	Hydrous titanium (IV) oxide added, 1.2 g $TiCl_4$ daily
25-31	1.45 x 97 (day 31)	230	Hydrous titanium (IV) oxide-cellulose chelate added, 1.2 g $TiCl_4$ + 1.2 g cellulose
56-75	1.40 x 90 (day 60)	236	Hydrous titanium (IV) oxide-cellulose chelate added, 1.2 g $TiCl_4$ + 1.2 g cellulose

titanium (IV) oxide have already been produced (7) and clearly
such are additionally advantageous where specialised restricted
movement of the immobilised cell particles is called for.   As
with immobilised enzymes, immobilised cell systems have been
applied to very few industrial processes, partially because of
the inertia of establishing methods and partially because of the
increased difficulty of operating such systems. It seems
certain, however, that the inherent advantages of these systems
will eventually prevail and there will be an increasing use of
immobilised systems by the pharmaceutical, chemical and food
industries.

Abstract

The use of immobilised cells for industrial and analytical
enzymic processes is prophetically advantageous, the problems
of isolation of the enzyme(s) and separation of the enzyme(s)
from the product being avoided. However, the majority of the
reaction currently used for direct enzyme immobilisation would
cause cell death if applied to cells. Our approach has been
based on the ability of water-insoluble metal hydroxides to
chelate and retain peptides, proteins, etc. including enzymes.
From various studies it was concluded that gelatinous titanium
and zirconium hydroxide matrices are effective matrices for
enzyme etc. immobilisation. Their advantages include low cost,
convenient preparation (which may be conducted in any
location without specialised facilities), the absence of any need
for pre-preparation, ability to couple enzyme at neutral pH, the
high retentions of specific activity of the enzyme on immobilisation, and the ability of modification to exert microenviron-
mental effects on and thereby alter the characteristics of the
immobilised enzyme.

Using this process, Saccharomyces cerevisiae and
Escherichia coli have been immobilised, and the continuation of
the living processes of the cells in the immobilised state were
demonstrated by oxygen uptake experiments. That the cells
become firmly bound to the support was demonstrated by
following the course of immobilised coloured cells (Serratia
marcescens), no colour being released during continuation of
the living processes of the cells.

The process of immobilisation on transition metal
hydroxides has now been extended to immobilisation of cells of
species of Acetobacter on hydrous titanium [IV] oxide and use
of the cells in this form for the continuous production of malt

vinegar from wort in tower fermenters. The efficiency of the fermenter was increased well above its normal maximum throughput using only free cells. With other species of Acetobacter, pre-chelation of the titanium species with cellulose further improved the efficiency of the fermenter.

Literature Cited

1. Kennedy, J. F., Chemically reactive derivatives of polysaccharides. Advances Carbohydrate Chem. Biochem., (1974) 29 305.

2. Kennedy, J. F., Macromolecules, Specialist Periodical Reports, Carbohydrate Chemistry, Part II, Vols. 4-11, The Chemical Society, London, 1971-1978.

3. Kennedy, J. F., Barker, S. A. and Humphreys, J. D., Insoluble complexes of amino acids, peptides and enzymes with metal hydroxides. J. Chem. Soc., Perkin I, (1976) 962.

4. Kennedy, J. F. and Kay, I. M., Hydrous titanium oxides-new supports for the simple immobilisation of enzymes. J. Chem. Soc., Perkin, I, (1976) 329.

5. Kennedy, J. F., Barker, S. A. and White, C. A., Immobilization of a-amylase on polyaromatic and titanium compounds incorporating a magnetic material. Die Stärke, (1977) 29 240.

6. Kennedy, J. F. and Humphreys, J. D., Active immobilized antibiotics based on metal hydroxides. Antimicrobial Agents and Chemotherapy, (1976) 9 766.

7. Kennedy, J. F., Barker, S. A. and White, C. A., The adsorption of D-glucose and glucans by magnetic cellulosic and other magnetic forms of hydrous titanium(IV) oxide. Carbohydrate Res., (1977) 54 1.

8. Frings GmbH, British Pat. 1, 101, 560 (1968).

9. Royston, M. G., Tower fermentation of beer. Process Biochem., (1966) 1 215.

10. Shore, D. T. and Royston, M. G., Chemical engineering of the continuous brewing process. Chem. Eng. (London), (1968) 46 No. 218, CE99.

RECEIVED February 15, 1979.

Enzymatic Synthesis of Pantothenic Acid by *Escherichia coli* Cells

Y. KAWABATA and A. L. DEMAIN

Department of Nutrition and Food Science, Massachusetts Institute of Technology, Cambridge, MA 02139

Pantothenic acid is both a component of coenzyme A and a vitamin of importance in human and animal nutrition. At present, it is produced commercially by chemical synthesis. Pantothenic acid is produced naturally by most microbial species, its formation being catalyzed by pantothenic acid synthetase (EC.6.3.2.1) (1,2). The enzyme-catalyzed reaction is as follows:

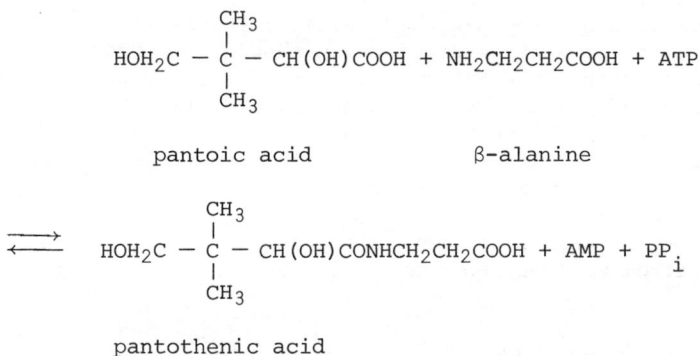

$$HOH_2C - \overset{\overset{\displaystyle CH_3}{|}}{\underset{\underset{\displaystyle CH_3}{|}}{C}} - CH(OH)COOH + NH_2CH_2CH_2COOH + ATP$$

pantoic acid   β-alanine

$$\underset{\longleftarrow}{\overset{\longrightarrow}{}} \quad HOH_2C - \overset{\overset{\displaystyle CH_3}{|}}{\underset{\underset{\displaystyle CH_3}{|}}{C}} - CH(OH)CONHCH_2CH_2COOH + AMP + PP_i$$

pantothenic acid

Since pantothenic acid contains an asymmetric carbon atom, chemical synthesis yields the racemic mixture. On the other hand, when enzymatic synthesis is used, only the biologically active form of pantothenic acid is produced.

In the present work, the use of Escherichia coli cells, instead of enzyme preparations, as the source of pantothenic acid synthetase was studied. Success was achieved in the production of the vitamin by frozen-thawed cells.

Pantothenate Production by Cells of E. coli ATCC 9637

E. coli ATCC 9637 was chosen for our studies since a considerable amount of work has been done on the pantothenic acid synthetase of this organism (3,4,5). The culture was grown in a

0-8412-0508-6/79/47-106-133$05.00/0
© 1979 American Chemical Society

medium containing glucose, ammonium lactate, enzyme- or acid-hydrolyzed casein and yeast extract (all at 2 g/liter) plus mineral salts. Assay of the enzyme activity was carried out in a 2-ml volume containing 40 μmoles β-alanine, 40 μmoles potassium pantoate, 20 μmoles disodium ATP, 200 μmoles KCl, 20 μmoles $MgSO_4$, 70 μmoles Tris·HCl buffer and 0.2 ml of an enzyme source. Pantothenate was determined by a disk-agar diffusion assay (6) using Lactobacillus plantarum ATCC 8014.

In an early experiment, we learned that the enzyme was produced during the exponential growth phase and declined in activity thereafter. This experiment was done using cells as enzyme source, since resting cells of E. coli had been reported (3) to produce and excrete pantothenic acid. Since our cells were washed twice with buffered saline and stored overnight in the refrigerator before being tested, it was reasonable to assume that their permeability properties could have been altered. This assumption was confirmed by the observation that such cells were markedly hindered in their ability to produce pantothenate when ATP was omitted from the reaction mixture. This dependency on exogenous ATP suggested to us that cell preparations treated even more drastically would be more active in pantothenate production. Indeed, we found that freezing and thawing the cells markedly increased activity.

A comparison of frozen-thawed cells, acetone-dried cells and a buffer extract of acetone-dried cells showed the order of decreasing activity (on a volumetric broth basis) to be acetone-dried cells > buffer extract > frozen-thawed cells. Despite this observation, we decided to use frozen-thawed cells for our further studies for the following reasons: a) the preparation of such an enzyme source is very simple and convenient; b) if the process eventually assumes industrial importance, the large amounts of acetone and ether needed to prepare acetone-dried cells or extracts therefrom would be both expensive and a waste-treatment problem.

Properties of Frozen-thawed Cells

Using frozen-thawed cells, we next determined the importance of each component of the reaction mixture. Omission of β-alanine, K-pantoate, ATP, $MgSO_4$ or enzyme eliminated virtually all activity. The product found in the extracellular fluid in the presence of all components was pantothenate, as identified by bioautography on paper.

Although permeability modification is certainly important for producing pantothenate by cell preparations, the alteration produced by freezing and thawing must be rather subtle, since pantothenic acid synthetase was found not to leak out of the frozen-thawed cells (Figure 1). This experiment was done by allowing a reaction to proceed for 20 hours, at which time the cells were removed by centrifugation. Little to no pantothenate was produced during the next 28 hours in the presence of the supernatant

fluid. However, in a control reaction mixture that was not centrifuged, pantothenic acid was produced in a linear fashion
throughout the 48-hour incubation period.

The optimum pH for pantothenate formation by frozen-thawed
cells was found to be 8.0 and the optimum temperature about 50°C.
Maas (3) reported that extracts of acetone-dried cells were
optimal at pH 8.5 to 9.0, and resting cells at 7.0 to 7.5.
Miyatake et al. (4,5) found the optimum for pure enzyme to be 10.0.
Frozen-thawed cells apparently behave with respect to pH in an
intermediate manner reflecting their nature as being neither cell-
free nor intact.

Of importance is the finding of Maas (3) that extracts of
acetone-dried cells have an optimum temperature of 25°C, are
stable at 35°C, but lose activity after one hour at 45°C.
Miyatake et al. (4,5) found 30°C to be optimal for pure enzyme
and a 10-minute incubation at 60°C to inactivate the enzyme almost
completely. In contrast to the reported lability of extracts and
pure enzymes, our optimum temperature of 50°C for frozen-thawed
cells was determined in a prolonged experiment that lasted 24
hours. It is thus obvious that frozen-thawed cells have a
stability advantage over both extracts and pure enzymes.

Immobilization of Frozen-thawed Cells

Immobilization was studied as a potential means to increase
stability further and to facilitate re-use of the preparation.
Agar was chosen as the support to be tested because of convenience
and the mild conditions necessary for immobilization. A 3% agar
solution at 40-50°C was mixed with a concentrated suspension of
frozen-thawed cells held at 30-40°C. The suspension was poured
into 1-ml aluminum cups and allowed to solidify. A cylinder of
agar (12 mm diameter x 10 mm) was removed from each cup and used
as the enzyme source. The enzyme activity of this immobilized
preparation is shown in Figure 2; the requirement for ATP is
easily seen. Such preparations have a half-life of about 100
hours at 37°C when incubated in the presence of reaction mixture.
About 1000 hours are necessary to inactivate such a preparation
completely.

Discussion

Although it was known almost 30 years ago (3,7) that resting
cells of E. coli could produce pantothenate from pantoate and
β-alanine, only in the present work was it found that cells respond to exogenous ATP. We were pleased to find that frozen-
thawed cells responded to an even greater degree, and that such
preparations were more stable than crude extracts of acetone-
dried cells or pure enzyme. These findings point to the use of
immobilized cells for the production of pantothenic acid and,
indeed, we found that agar-immobilized cells are active. It

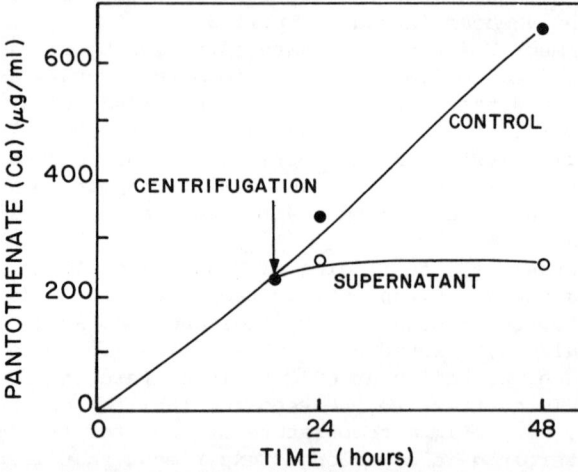

Figure 1. *Nonleakage of pantothenic acid synthetase from frozen–thawed cells. After 20 hr of reaction, the cells were removed from the experimental reaction*

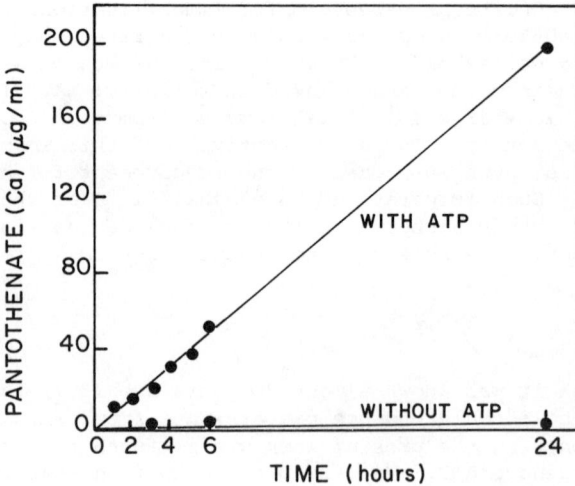

Figure 2. *Effect of ATP on pantothenic acid synthetase activity of frozen–thawed cells immobilized in agar. The experiment was carried out at 30°C and pH 8.0*

should be noted that a positive response to ATP by cellular preparations has also been observed in other systems (8,9,10,11,12).

Abstract

Pantothenic acid, a vitamin of importance in human and animal nutrition, is produced commercially by chemical synthesis. We investigated its bioconversion from β-alanine, potassium pantoate and ATP by various types of Escherichia coli ATCC 9637 cell preparations in Tris buffer containing KCl and $MgSO_4$. We found exponential phase cells to be most capable of carrying out this reaction. ATP had a marked stimulatory effect on the reaction even though cells were used. Pantothenate production was also observed with acetone-dried cells and crude extracts. A marked increase in pantothenate production by cells was effected by freezing and thawing. No enzyme activity leaked from the cells during 20 hours of reaction. Production by frozen-thawed cells was considerably more stable to heat than that reported for crude extracts or pure enzyme. Immobilization of frozen-thawed cells in agar yielded active preparations, which required 1000 hours at 37°C for complete inactivation.

Acknowledgment: The authors are indebted to Toray Industries, Inc., for financial support of Y. K. during his stay at M. I. T. Y. K.'s permanent address is: Basic Research Laboratories, Toray Industries, Inc., Kamakura, Japan.

Literature Cited

1. Brown, G.M. Biosynthesis of pantothenic acid and coenzyme A. In: "Comprehensive Biochemistry," vol. 21 (M. Florkin and E.H. Stotz, eds.), p. 75. Elsevier, Amsterdam, 1971.
2. Matsuyama, A. Bull. Agr. Chem. Soc. (Japan) (1957) 21, 47.
3. Maas, W.K. J. Biol. Chem. (1952) 198, 23.
4. Miyatake, K., Nakano, Y. and Kitaoka, S. Agr. Biol. Chem. (1973) 37, 1205.
5. Miyatake, K., Nakano, Y. and Kitaoka, S. J. Nutr. Sci. Vitaminol. (1978) 24, 243.
6. Kojima, H., Matsuya, Y., Ozawa, H., Konno, M. and Uemura, T. J. Agr. Chem. Soc. (Japan) (1958) 32, 33.
7. Maas, W.K. J. Bacteriol. (1950) 60, 734.
8. Ogata, K., Shimizu, S. and Tani, Y. Agr. Biol. Chem. (1970) 34, 1757.
9. Shimizu, S., Miyata, K., Tani, Y. and Ogata, K. Biochim. Biophys. Acta (1972) 279, 583.
10. Shimizu, S., Tani, Y. and Ogata, K. Agr. Biol. Chem. (1972) 36, 370.
11. Shimizu, S., Morioka, H., Tani, Y. and Ogata, K. J. Ferm. Technol. (1975) 53, 77.
12. Uchida, T., Watanabe, T., Kato, J. and Chibata, I. Biotech. Bioeng. (1978) 20, 255.

RECEIVED February 15, 1979.

The Use of Whole Cell Immobilization for the Production of Glucose Isomerase

STEPHEN J. BUNGARD, ROGER REAGAN, PETER J. RODGERS, and KEVIN R. WYNCOLL

ICI Agricultural Division, Billingham, Cleveland, United Kingdom

The subject of this paper is the use of a whole cell immobilisation technique for the production of an immobilised glucose isomerase enzyme system and the subsequent use of this enzyme in the production of high fructose corn syrups (see Fig. 1). The HFCS industry has undergone dramatic growth in the 1970's and is predicted to maintain a high growth rate into the 1980's. The development of this industry has been made possible by the discovery of an enzyme, glucose isomerase, capable of catalysing the transformation of glucose to fructose, and the development of commercially viable enzyme immobilisation procedures. This paper deals with the technology of the production and properties of one such system.

Production Process

Figure 2 shows the production process to be divided into the following stages.

1. The fermentation of a culture of a bacterium which produces glucose isomerase.
2. A flocculation process. This serves to harvest the bacterial cells, i.e. remove cells from the growth medium, and ultimately to form the support matrix. This stage in the process is achieved by the sequential addition of cationic and anionic polyelectrolytes. The two polyelectrolytes used are capable of forming a polysalt complex and it is believed that this complex structure, in which the bacterial cells are embedded, gives rise to the product's robust physical characteristics whilst allowing the desired enzyme performance.
3. Product forming. The remainder of the process is one of water removal which is necessary for microbiological stability and ease of transportation, and product shaping. The dry product is roughly cylindrical with approximate size dimensions of 1.2mm diameter and 2mm length.

0-8412-0508-6/79/47-106-139$05.00/0

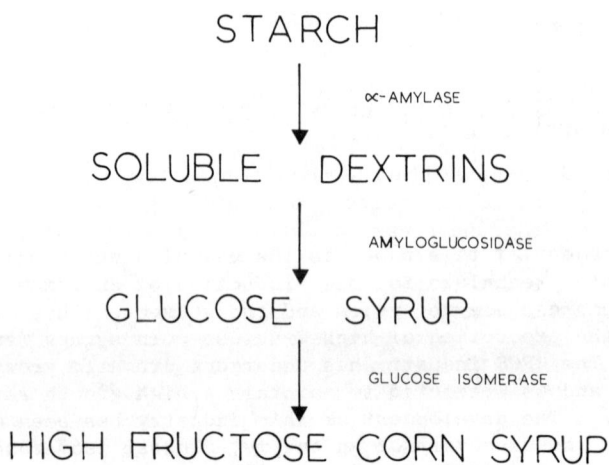

Figure 1.   *The conversion of starch to high fructose corn syrup*

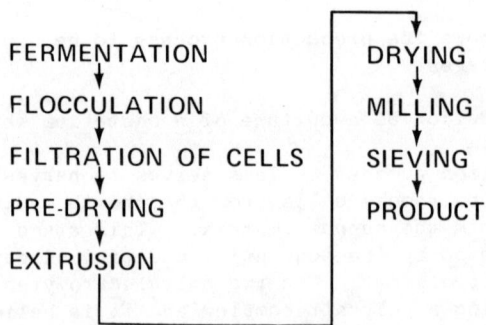

Figure 2.   *Production process for glucose isomerase*

Properties of the Product

In use the pellet takes up water and swells to approximately three times its dry volume. Electron microscope studies (see Fig. 3) indicate a porous honeycomb structure in which the individual bacterial cells are homogeneously distributed throughout the pellet. The relevance of the pore dimensions will be considered later.

In commercial application the enzyme is packed into a fixed bed column reactor through which is passed a concentrated (45% w/w solids) dextrose solution. During passage through the column, isomerisation occurs resulting in a syrup typically containing 42% DS fructose. The conditions of operation are such that the temperature is controlled in the region of 60°C and a pH of 8.0. Under these conditions, the typical life curve has the form shown in Figure 4. By the time the production rate has reached 25% of the initial value the enzyme has produced over 3,000 lb of high fructose corn syrup (dry basis) per lb weight of enzyme (when using a pure dextrose feed). Careful examination of this curve reveals a linear phase followed by an exponential phase. At start-up of the column, the production rate does not conform to this pattern until steady state conditions are established.

Diffusion-Controlled Rate Model

A mathematical model has been developed which described the properties of the immobilised enzyme system and which gives an explanation of the previously mentioned behaviour. Typically enzymes exhibit exponential loss of activity with time at elevated temperatures which is brought about by the process of thermal denaturation. Therefore a process of linear decay, as shown by this system, must require some other explanation. Many other solid phase catalysts used in the chemical industry are known to exhibit bulk diffusional limitation. That this should also apply to an immobilised enzyme preparation is a reasonable starting hypothesis for the development of a mathematical model of this system. As described by my colleague in a previous paper (ref 1) such a model has been shown to predict results closely similar to those obtained in practice. The linear portion of the curve occurs as a consequence of the diffusional limitation present within the pelleted structure; but as the activity of the glucose isomerase enzyme declines, diffusion of substrate and product within the pellet is no longer the rate limiting step and the expected exponential phase becomes apparent. The theoretically predicted diffusional limitation is substantiated by electron micrograph and micromeritic studies on samples with a wide range of bulk activities. That is to say, samples with similar intrinsic activities can display widely varying bulk activities due to differences in the porosity of the pellet causing differ-

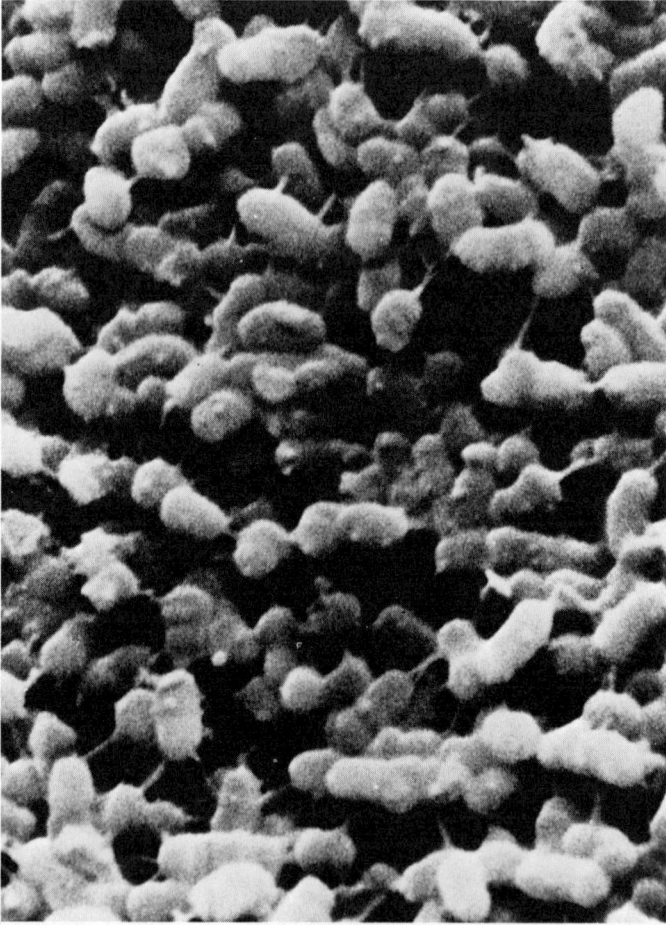

Figure 3.   Electron micrograph of a pellet of glucose isomerase

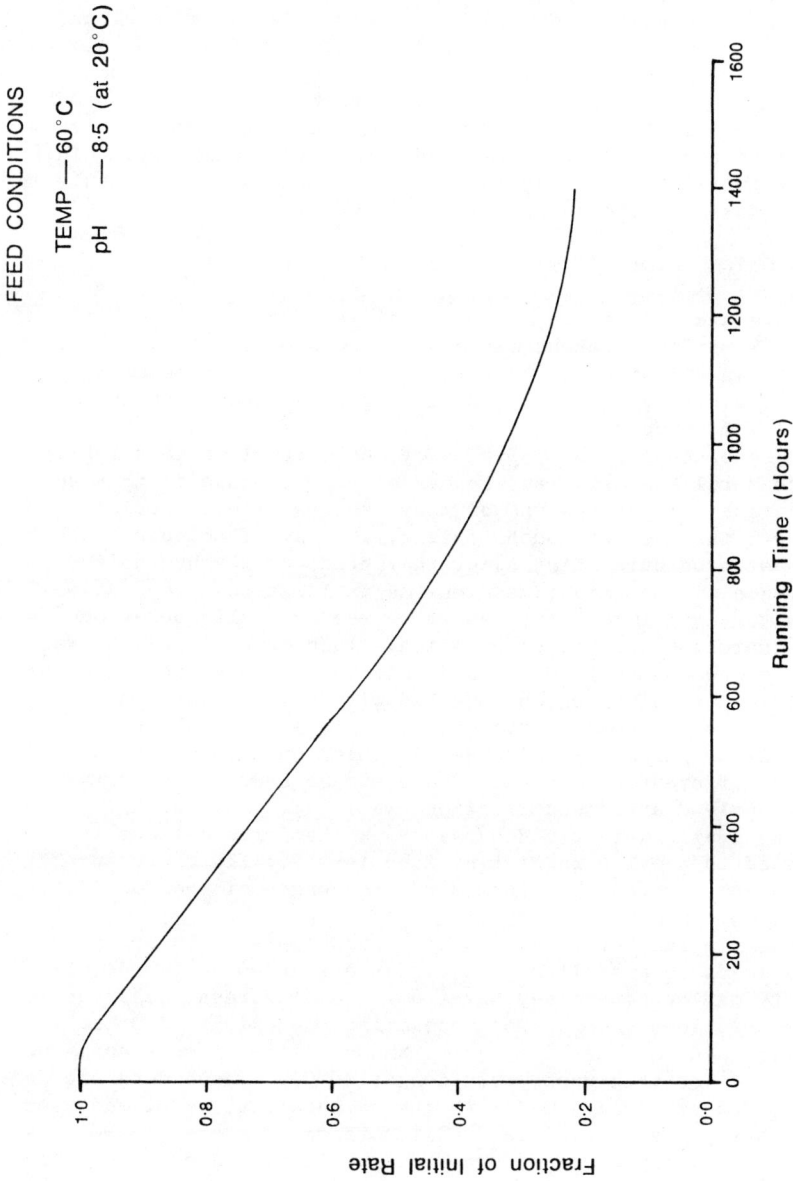

Figure 4. *ICI glucose isomerase characteristic life curve. Feed conditions: temperature 60°C, pH 8.5 (at 20°C).*

ences in the effect of diffusional limitation. Figure 5 shows
two very dissimilar internal structures which result from
different conditions during the processing steps, particularly
during the flocculation procedure. This model, when combined
with kinetic analysis of the enzyme reaction, allows not only an
understanding of the effect of processing conditions on the final
product activity to be gained but also enables predictions to be
made of the effect of various conditions, such as temperature,
pH, substrate concentration, etc on the enzyme behaviour. This
information is valuable for the operation of commercial plants in
order to maximise the potential of the enzyme.

Reactor Column Hydraulics

There are, of course, a number of ways in which the diffusional limitation may be minimised, the simplest of which is a
reduction in the size of the particle. However, an equally
important property of any immobilised glucose isomerase enzyme
system for commercial application, is the ability to withstand
applied load conditions. Typically, commercial columns are up
to 20 feet tall and six feet in diameter. Because of this scale
of operation, direct testing of many samples is not feasible.
To obviate the need for such testing, ICI have developed a laboratory test procedure which accurately predicts the hydraulic
performance of the immobilised enzyme in large-scale operation.

As for many immobilised enzymes, the hydraulic behaviour is
not adequately described by classical fluid mechanics. It was,
therefore, necessary to develop a detailed mathematical model of
the column hydraulics which together with a laboratory test
procedure, would provide data on the basic mechanical properties
of the enzyme pellet. The model is based on a force balance
across a differential element of the enzyme bed. The primary
forces involved are fluid friction, wall friction, solids
cohesion, static weight and buoyancy. The force balance is
integrated to provide generating functions for fluid pressure
drop and solid stress pressure down the length of the column
under given conditions.

The main laboratory test uses a porous piston to apply
various loads to a small sample of the enzyme so that bed permeability may be determined under a range of stress pressures in
glucose solutions under normal operating conditions. Other
tests are used to measure adhesive and cohesive effects and true
solid densities in glucose solution at $60^{\circ}C$. These data are fed
into a computer, programmed with the mathematical model and used
to plot out pressure profiles under a range of operating
conditions for any given column configuration. Because numerical
analysis is used, the method is completely flexible and does not
rely on forcing the data to fit a particular empirical correlation.

The predictions from the mathematical model have been tested
against the performance of several large scale columns, typically

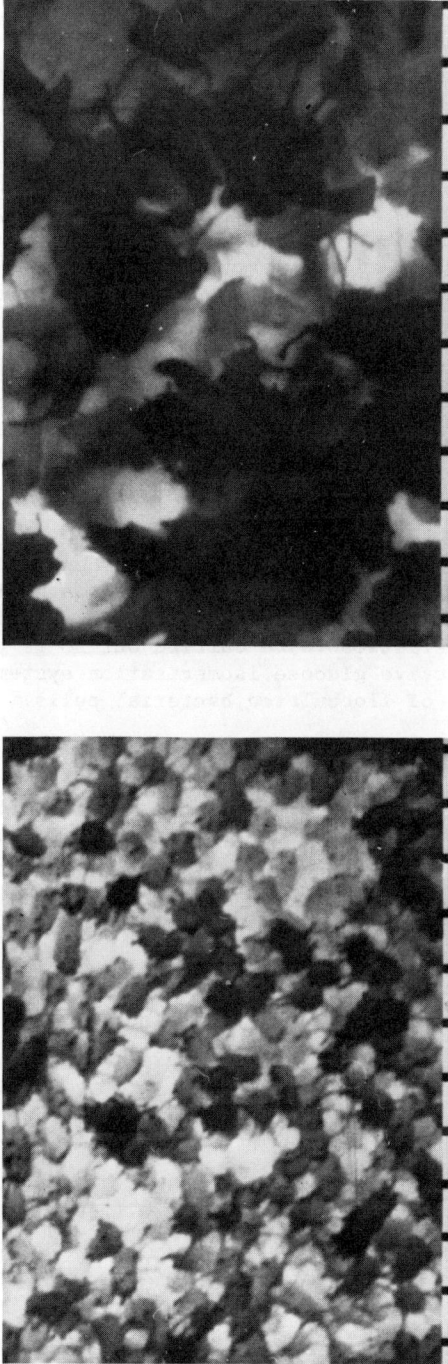

Figure 5. Internal structures of glucose isomerase pellets. Left: homogeneous structure—well distributed cells from a good flocculation. Right: heterogeneous structure—clustered cells from an earlier, poorer flocculation. Width of each photograph represents 1 μ.

with a bed depth of 15 feet.   Because of the degree of detail in
the model and the precision of the laboratory tests, agreement
between predicted and actual performance has been excellent.
    An abbreviated version of the test procedure, incorporating
an additional safety factor is used for routine product quality
control and for the screening of development products.   With
this test, a reliable prediction of the enzyme's column performance can be made from a small sample in about an hour.   It is
important to realise that the hydraulic properties of the enzyme
bed are related to the enzyme particle size distribution and
rigidity.   These properties in turn can be influenced by environmental factors, e.g. pH, ionic concentration, temperature and
these effects have been studied extensively.
    The ICI enzyme is characterised by rapid equilibration to a
high permeability under loaded conditions which results in a low
and stable pressure drop during column operation and hence beds
of enzyme can be operated with gravity feed.

Summary

    The combination of the two models previously discussed
provides a powerful tool in understanding the behaviour of the
immobilised enzyme system.   This enables the optimisation of
the enzyme characteristics to be carried out so as to produce a
commercially attractive glucose isomerisation system which is
based upon the use of flocculated bacterial cells.

Literature Cited

    Ref 1.   Reagan R.   1977.   Mathematical Modelling in
Enzyme Systems: Some Effects of Diffusional Limitation.   1977
International Biochemical Symposium, Toronto.

RECEIVED February 15, 1979.

Temperature Dependence of the Stability and the Activity of Immobilized Glucose Isomerase

J. A. ROELS and R. VAN TILBERG

Gist-Brocades N.V., Research and Development, P.O. Box 1,
2600 MA Delft, The Netherlands

A number of microorganisms are capable of transforming glucose into its isomer fructose by the action of the enzyme glucose-isomerase. This property is of potential commercial significance as the enzyme can in principle be used to produce a mixture of glucose and fructose using a corn based glucose syrup as a source of raw material. This mixture is termed high fructose corn syrup (HFCS). HFCS is considered to be an important competitor for saccharose as a sweetener.

In industrial practice an immobilized form of glucoseisomerase is used. Gist Brocades' immobilized glucoseisomerase, Maxazyme GI-immob consists of pasteurized whole cells of *Actinoplanes missouriensis* entrapped in gelatin in such a way that, after cross-linking with glutaraldehyde, the substrate, glucose, and the product, fructose, can diffuse more or less freely into and out of the particles.

In this paper a mathematical model will be presented describing the conversion process in a fixed bed reactor. The model allows the calculation of the temperature dependence of the initial acitivity of the immobilized enzyme. It also predicts the stability of that activity as a function of the operating temperature. The model is of an approximative nature and the simplifications which are introduced allow an analytical solution of the equations of the model. The results of the theoretical deductions are verified experimentally.

Mathematical model

Enzyme kinetics. The kinetics of transformation processes catalysed by a single enzyme are often described using the Michaelis-Menten equation (1). The derivation of this equation is, however, based on two assumptions. The pseudo steady state hypothesis (2) with respect to the intermediary enzyme-substrate complex is valid and the reverse reaction from product to substrate can be

0-8412-0508-6/79/47-106-147$06.50/0

neglected. In general the first assumption is taken for granted in enzyme catalysis. The latter assumption is only justified if the absolute value of the free enthalpy change of the reaction is large compared to the product of universal gas constant and absolute temperature. This presents problems in the case of the conversion of glucose to fructose, the standard free enthalpy change being -400 J/mole at 60°C; the product of universal gas constant and absolute temperature being 2750 J/mole.

Fratzke et al. (3) performed a mathematical analysis based on the following kinetic scheme:

$$G + E \underset{k_{-1}}{\overset{k_1}{\rightleftharpoons}} XE \underset{k_{-2}}{\overset{k_2}{\rightleftharpoons}} E + F \tag{1}$$

in which G   is glucose
       E   is the free enzyme
       XE   is the intermediary enzyme-substrate complex
       F   is fructose
       k_1, k_{-1}, k_2, k_{-2} are kinetic constants

As can be seen the reverse reaction is included in the scheme proposed by Fratzke et al. (3).

The results of their analysis, which again includes the steady state hypothesis with respect to the enzyme-substrate complex, is the following kinetic equation:

$$r_s = \frac{r'_{s, max} (C_s - C_s^x)}{K'_s + (C_s - C_s^x)} \tag{2}$$

in which r_s is the rate of conversion of glucose to fructose (mole/m^3s)
       $r'_{s, max}$ is the apparent maximal forward rate of the conversion of glucose to fructose (mole/m^3s)
       K'_s is a pseudo Michaelis-Menten constant for the substrate (mole/m^3)
       C_s is the glucose concentration (mole/m^3)
       C_s^x is the glucose concentration corresponding to thermodynamic equilibrium (mole/m^3)

If equation (2) is applied to a conversion process using the free whole cells of an organism a convenient formulation of equation (2) is the following:

$$r_s = \frac{V'_{s, max} \cdot C_E (C_s - C_s^x)}{K'_s + (C_s - C_s^x)} \tag{3}$$

in which $V'_{s, max}$ is the maximum specific rate of fructose formation (mole/kg organism dry matter ss)

C_E is the concentration of the organism (kg dry matter/m^3)

The "constants" $V'_{s, max}$ and K'_s are rather complex functions of the kinetic constants and the equilibrium glucose concentration:

$$V'_{s, max} = \frac{K^x + 1}{K^x} \cdot \frac{K_{MF}}{K_{MF} - K_{MG}} \cdot k_2 [E] \qquad (4)$$

$$K'_s = \frac{1}{K_{MF} - K_{MG}} \{(K_{MF} + K_{MG} \cdot K^x) \cdot c_s^x + K_{MF} \cdot K_{MG}\} \quad (5)$$

in which K^x is the equilibrium constant for the conversion of glucose to fructose (-)

K_{MF} is the Michaelis-Menten constant for the conversion of fructose to glucose, being equal to $(k_{-1} + k_2) / k_{-2}$ (mole/m^3)

K_{MG} is the Michaelis-Menten constant for the conversion of glucose to fructose, being equal to $(k_{-1} + k_2) / k_1$ (mole/m^3)

$[E]$ is the intrinsic enzyme concentration per unit of microorganism dry matter (mole/kg dry matter)

Adopting the values of the kinetic constants given by Fratzke et al. it can be shown that for the conditions prevailing in the fixed bed conversion process, $C_s - c_s^x$ is small as compared to K'_s. At an initial syrup glucose concentration of 3000 moles/m^3 and a relative conversion to fructose of 45%, K'_s is of the order of 5000 moles/m^3 and $C_s - c_s^x$ varies between about 1000 moles/m^3 and 20 moles/m^3 (column inlet and column outlet respectively).

Under these conditions equation (3) can to a fair degree of approximation be simplified to:

$$r_s = K (C_s - c_s^x) \qquad (6)$$

in which K is a pseudo first order rate constant given by:

$$K = \frac{k_2 (K^x + 1) [E] \cdot C_E}{K^x \left[(1 + \frac{K_{MG}}{K_{MF}} K^x) c_s^x + K_{MG}\right]} \qquad (7)$$

It is important to note that equation (7) implies the pseudo first order rate constant to be a function of c_s^x and hence of the initial glucose concentration C_{s0}, K formally cannot be considered a true kinetic constant. For the purpose of the present model, describing a situation in which the initial glucose concentration is a constant, K can be considered to be a constant but if the results are translated to other initial glucose concentrations equation (7) has to be taken into account.

The efficiency factor for an immobilized enzyme. In general the conversion rate of an immobilized enzyme is lower than that of an equal amount of the free enzyme. This decreased activity is caused by diffusional limitations to the rate at which the subtrate is transported to the site of reaction in the immobilized enzyme particles. In chemical engineering the subject of the interplay between diffusional limitations and chemical kinetics in heterogeneous catalysis has been extensively studied. The state of the art on this subject is described by Satterfield (4).

For the case of a first order reaction in a spherical particle a relationship between an efficiency factor, η, and a dimensionless number, the so called Thiele factor, Φ, can be shown to be given by:

$$\eta = \frac{3}{\Phi} \left(\frac{1}{\tanh\Phi} - \frac{1}{\Phi} \right) \tag{8}$$

in which η is the ratio of the general conversion in the particle to the conversion in absence of diffusional limitations (-)
Φ is the Thiele factor being defined as

$$\Phi = R_p \sqrt{\frac{k}{D}} \tag{9}$$

in which k is a first order reaction rate constant (1/s)
D is the diffusivity of the reactant in the catalyst particle (m^2/s)
R_p the particle radius (m)

If the simplified pseudo first order equation for the conversion rate of glucose to fructose, equation (6), is assumed to be sufficiently accurate, the conversion rate of particles in which whole cells are immobilized is given by:

$$r_s = K \eta C_{si} V_e \tag{10}$$

in which η is the efficiency factor given by equation (8), the Thiele factor being given by

$$\Phi = R_p \sqrt{\frac{K}{D}} \tag{11}$$

K is the pseudo first order rate constant given by equation (7) (1/s)
D is the coefficient of diffusion for glucose in the particles (m^2/s)

C_{si} is the substrate concentration at the particles interface (mole/m^3)

V_e is the amount of immobilized enzyme present (m^3)

The effect of the temperature. The operating temperature
affects the isomerization process in two important ways.

Firstly the pseudo first order rate constant K is expected
to increase with increasing temperature. In the present treatment
the relationship between K and temperature will be assumed to be
of the Arrhenius type:

$$K = A_1 e^{-\Delta H_1^{\text{x}} / RT} \tag{12}$$

in which: A_1 is a constant (1/s)
ΔH_1^{x} is the activation enthalpy of the enzyme-catalysed-
isomerization of glucose to fructose (J/mole)
R   is the universal gas constant (J/mole K)
T   is the absolute temperature (K)

Secondly the deactivation rate of the enzyme activity is assumed to increase with increasing temperature. It is assumed that
the pseudo first order rate constant K decreases with time according to:

$$K = K_o e^{-k_d t} \tag{13}$$

in which K_o is the initial pseudo first order rate constant (1/s)
k_d is the deactivation constant (1/day)
t is the operating time (days)

For the temperature dependence of the deactivation constant
an Arrhenius relationship is assumed:

$$k_d = A_2 e^{-\Delta H_2^{\text{x}} / RT} \tag{14}$$

in which A_2 is a constant
ΔH_2^{x} is the activation enthalpy of the enzyme deactivation process

Model for fixed bed isomerization. In industrial practice
immobilized glucoseisomerase is often applied to the isomerization
of glucose in a fixed bed reactor.

Under the assumptions about the kinetics presented above the
construction of a simple mathematical model for this process is
quite straightforward. A balance equation for glucose over an in-
finitesimal slice of the fixed bed (see figure 1) can be formulated as follows:

$$V \, dC_s = -K \{c_s - c_s^{\text{x}}\} \cdot \eta \, (1 - \varepsilon) \, A \, dh \tag{15}$$

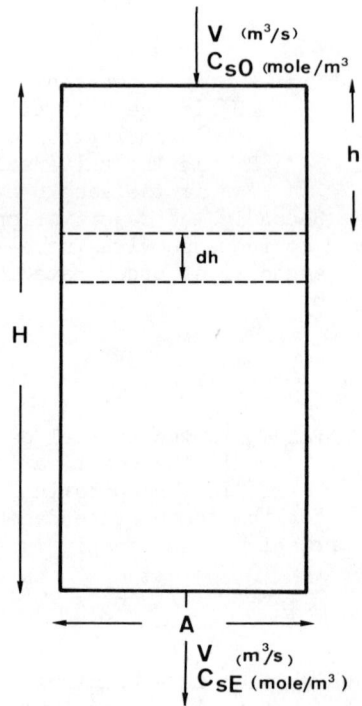

*Figure 1. Model for fixed-bed isomeri-
zation*

in which V is the flow rate of the glucose syrup through the
    fixed bed (m^3/s)
    ε is the porosity of the bed (-)
    A is the fixed bed cross section (m^2)
    h is the height coordinate (m)
The boundary conditions for equation (15) are:

$$C_s = C_{s0} \qquad h = 0$$
$$C_s = C_{sE} \qquad h = H \tag{16}$$

in which C_{s0} is the glucose concentration in the syrup entering
    the column ($mole/m^3$)
    C_{sE} is the glucose concentration in the isomerized syrup
    leaving the column ($mole/m^3$)
The solution of (15) with the boundary conditions according to
(16) is given by:

$$V = \frac{K \, \eta (1 - \varepsilon) \, A.H}{\ln ((C_{so} - C_s^x) / (C_{sE} - C_s^x))} \tag{17}$$

Several assumptions are implicit in the derivation presented here;
two of the important ones are:
- the resistance for mass transfer to the particles can be neglected
- the syrup flows through the column in ideal plug flow
    Equation (17) gives the relationship between the flow rate, V,
which results in conversion to an exit glucose concentration C_{sE}
as a function of K and η. As K was assumed to decrease with operating time (equation (13)) and η is, through the Thiele factor,
a function of K, the product $K\eta$ will decrease with time. The derivation presented above is only valid if the pseudo steady state hypothesis is invoked with respect to the change of K with
time. K is assumed to be virtually constant during the residence
time of the syrup in the column. If this assumption is not made
the simple differential equation (15) has to be replaced by a set
of simultaneous partial differential equations in time and height
and the solution becomes by no means trivial. The pseudo steady
state hypothesis with respect to K is, however, quite reasonable
as the residence time in the column is of the order of one hour
and the time constant of the deactivation process is of the order
of 1 to 100 days. Equation (17) can thus be used to specify the
flow velocity at any moment during the operating time if the momentary value of K is calculated according to equation (13).

Experimental

     The parameters of the model and their temperature dependences
were estimated on the basis of fixed bed experiments on a laboratory scale. The enzyme used was Gist-Brocades' immobilized glucose
isomerase, Maxazyme   GI-immob. The glucose "syrup" was a 45% w/w
solution of crystalline dextrose in distilled water containing
3 mM of $MgSO_4$ and 100 ppm SO_2. The pH of the syrup was 7.5. The
fixed bed enzyme reactor was a jacketed glass column provided with
a filter plate and containing about 30 ml of bulk volume of the
immobilized enzyme. The conversion was determined polarometrically
as well as by means of high performance liquid chromotography. The
range of operating temperatures was 55 to 75°C. Two types of experiments were performed. Three sets of experiments were performed
at constant flow rate of the substrate solution, the percentage
glucose isomerized being determined as a function of operating
time. One set of experiments was conducted at constant fractional
conversion to fructose, the flow rate being determined as a function of operating time. The duration of the experiments varied
from 5 days at the highest temperature to more than 100 days at
the lowest temperature.
     The parameters of the model were estimated from the experi-
mental data using a non linear multivariate curve fitting technique. In this process the temperature dependence of the diffusion
coefficient for glucose was assumed to be small in the range of
temperatures studied. The equilibrium constant K^x was assumed to
be given by:

$$K^x = 28.8 \exp (-1100 / RT) \tag{18}$$

This equation is based on data reported by Fratzke et al. (3).
     The diffusion coefficient for glucose in the immobilized en-
zyme particles, $\mathbb{D}$, was estimated to be 6.7×10^{-11} (m^2/s). A reasonable value if it is compared with the value of 8.8×10^{-11} (m^2/s)
obtained by Vellenga (5) for a different kind of immobilized glucoseisomerase.
     The dependence of the estimated values of the initial value
of the pseudo first order rate constant, K_o, on temperature was
interpreted in terms of an Arrhenius relationship. In figure 2 the
Arrhenius plot for K_o, standardized, within each set of experiments, with respect to the initial value of the pseudo first order
rate constant at 65°C, $K_{0,65}$, is shown. The activation enthalpy of
the reaction is estimated to be 79×10^3 J/mole. This can be compared with a value reported by Fratzke et al. being 70×10^3 J/mole.
     The dependence of the experimental values of the deactivation
constant on temperature is shown in an Arrhenius plot in figure 3.
The activation enthalpy of the deactivation reaction is estimated
to be 203×10^3 J/mole. This agrees well with earlier results of
Fratzke et al. (3), 204×10^3 J/mole and Nielsen (6), 197×10^3 J/mole.

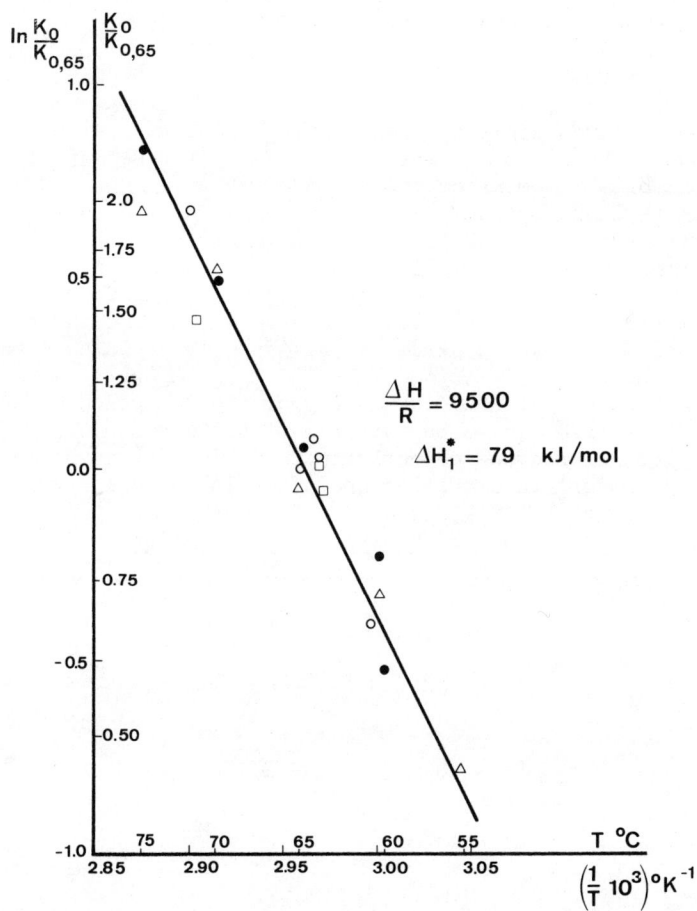

Figure 2.   Arrhenius relationship for fixed-bed initial activity. (●) Constant conversion, (△, □, ○) constant flow.

From figures 2 and 3 it is also clear that the results of
the constant flow experiments agree well with those of the constant conversion experiment. This may be considered an indication
that the assumption of negligable diffusional resistance for glucose transport to the particles is correct.

In table I the kinetic constants and their temperatures de-
pendences for the present immobilized enzyme and some other relevant data have been summarized.

Evaluation of the model

The model was applied to estimate the theoretical relationship between the fixed bed initial flow velocity resulting in 45%
of glucose being isomerized to fructose and the operating temperature for an immobilized enzyme defined by the characteristics given in table I. The results of this evaluation, in
terms of the initial flow velocity relative to that at $65^{\circ}C$, are
shown in an Arrhenius plot in figure 4. The experimental results
used in the parameter estimation are also shown. The theoretical
relationship is shown to be definitely non linear in an Arrhenius
plot.

To obtain a better understanding of the influence of the activity of the free enzyme on this relationship calculations were
performed for a free enzyme activity twice as well as half of
that of the reference situation specified in table I.

Table I. Parameter values for reference situation

$K_{0,65} = 3.0 \times 10^{-3}$		$(1/s)$
$K_{0,T} = 4.83 \times 10^{9} \exp\left(-\dfrac{9500}{T}\right)$		$(1/s)$
$k_d = 1.51 \times 10^{30} \exp\left(-\dfrac{24400}{T}\right)$		$(1/day)$
$D = 6.7 \times 10^{-11}$		(m^2/s)
$R_p = 6 \times 10^{-4}$		(m)

The resulting Arrhenius plots are shown in figure 5. As can be
seen the initial free enzyme activity affects the Arrhenius plot.
This, in combination with the already mentioned non linearity of
the Arrhenius plot, shows that direct estimation of the activation
enthalpy of reaction from an Arrhenius plot of the initial flow velocity cannot be right. This is also illustrated in figure 6; the

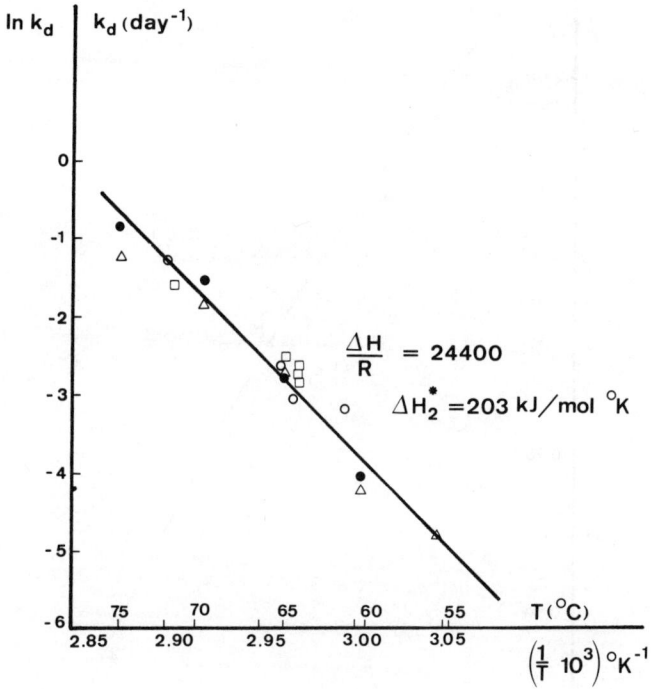

Figure 3.   Arrhenius relationship for deactivation constant. (●) Constant conversion, (△, □, ○) constant flow.

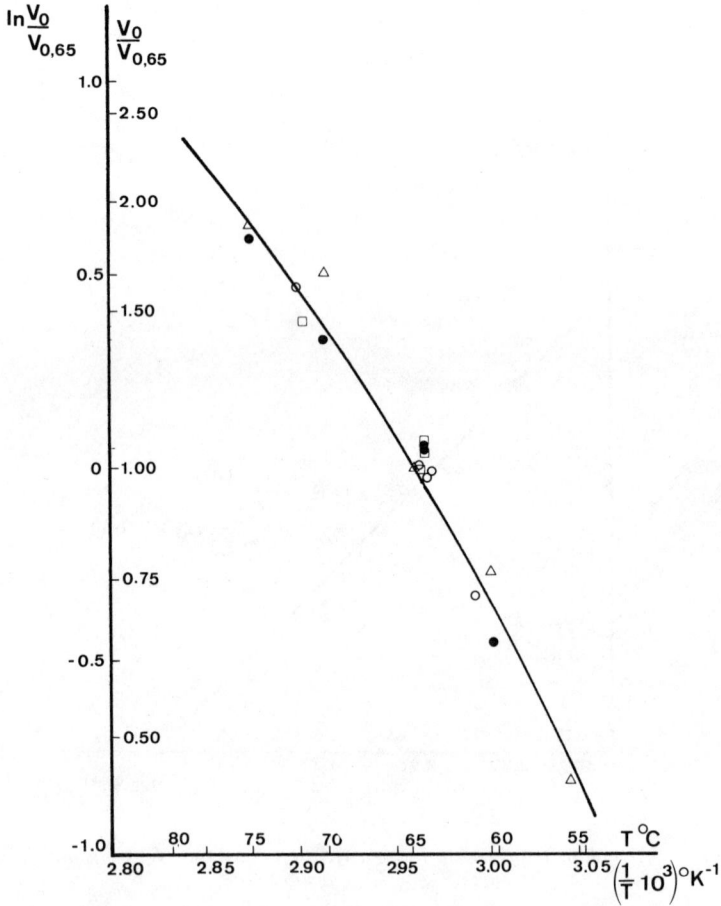

Figure 4. Experimental data and theoretical relationship between fixed-bed initial flow velocity and temperature. (●) Constant conversion, (△, □, ○) constant flow.

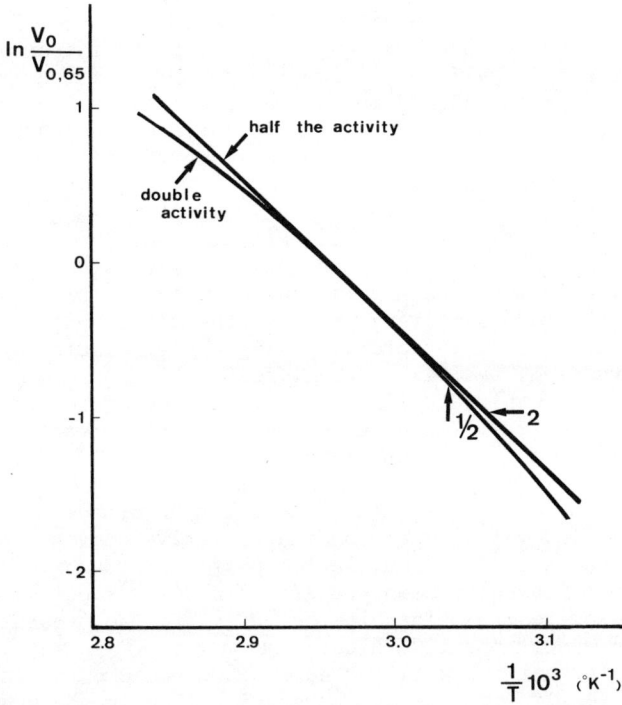

Figure 5.   Characteristics of the Arrhenius plot of fixed-bed initial flow velocity for different activities of the free enzyme

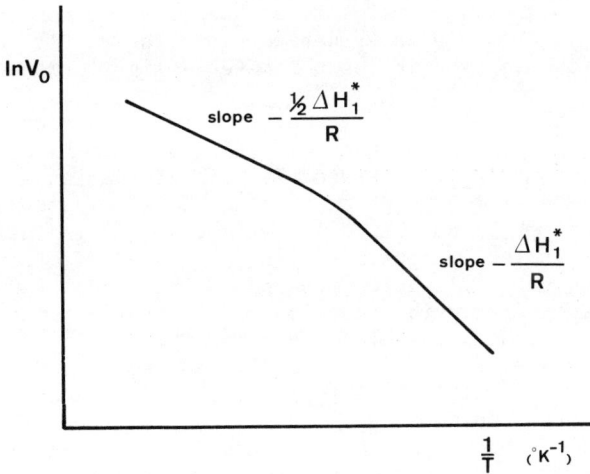

Figure 6.   Overall characteristics of the Arrhenius plot of fixed-bed initial flow velocity

characteristics of the Arrhenius plot of initial flow velocity are
shown for a broad temperature range. The slope of the Arrhenius
plot corresponds to half the activation enthalpy of the activity
of the free enzyme at high temperatures and is equal to the slope
corresponding to the activation enthalpy at low temperatures. In
the intermediary temperature range the slope gradually decreases
with increasing temperature.

In figure 7 the theoretical relationship between the initial
flow velocity resulting in a fixed percentage of conversion and
the activity of the free enzyme is shown. The relationship is linear if the free enzyme has a very low activity or, alternatively,
if the particle radius is very small or the diffusivity of glucose very high (low Thiele factor). At high Thiele factors (enzyme
activity very high, radius large, diffusivity low) a square root
relationship is obtained. To show the relevance of both types of
behaviour to the immobilized enzyme used in the present investigation its expected behaviour at temperatures of 50 and 80°C is
shown in figure 7.

The present model also allows calculation of the relationship between flow velocity at constant relative conversion and
operating time in a fixed bed. In the past linear as well as exponential functions have been proposed for this property. In figure 8 the expected relationships according to the present model,
linear decay and exponential decay are shown. The results
of one representative experiment have also been shown. The present theory results in a decay curve which is between the linear
and exponential relationships. The assumption of linear decay involves an underestimation of the activity half life; the assumption of exponential decay, however, overestimates the half life
of the initial flow velocity.

In figure 9 the theoretical relationship between the half
life of the initial flow velocity at 45% relative conversion and
temperature is shown in an Arrhenius plot. The results obtained
in the experiments used in the parameter estimation are also
shown.

The overall characteristics of the Arrhenius plot of flow velocity half life are shown in figure 10. The plot is linear with
a slope corresponding to the deactivation activation enthalpy of
the free enzyme at low temperatures (low Thiele factors). The half
life of the immobilized enzyme activity is equal to that of the
free enzyme. In the intermediary range of Thiele factors or temperatures the slope slowly decreases then increases again to the
slope corresponding to the deactivation activation enthalpy of
the free enzyme. At high Thiele factors or high temperatures the
slope again corresponds to the deactivation activation enthalpy
but the activity half life of the immobilized enzyme is twice
that of the free enzyme.

From the foregoing it is clear that the initial flow velocity half life of the immobilized enzyme is determined not only by
the deactivation properties of the free enzyme but also by the

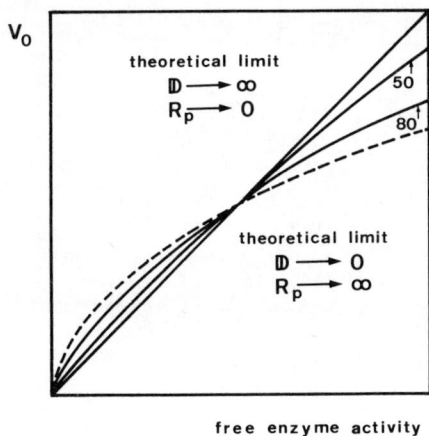

Figure 7. Relationship between fixed-bed initial flow velocity and free enzyme activity

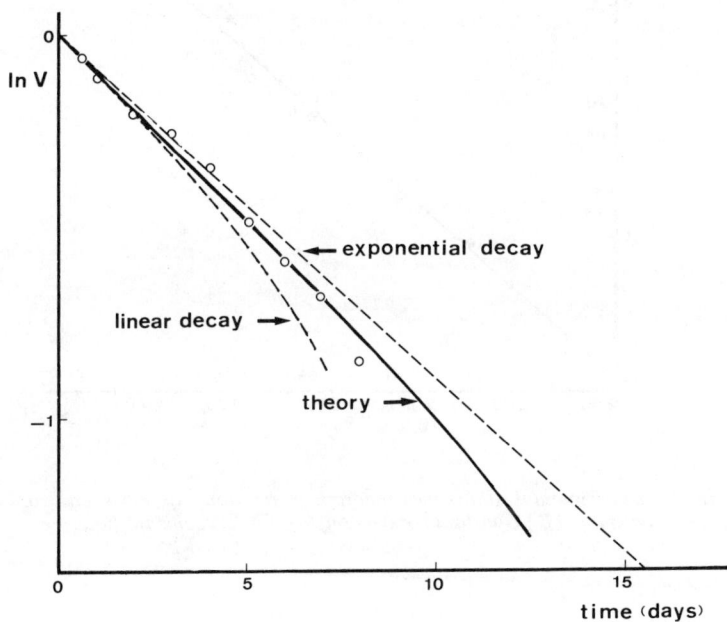

Figure 8. Decay of the fixed-bed velocity at constant relative conversion: (○) experimental.

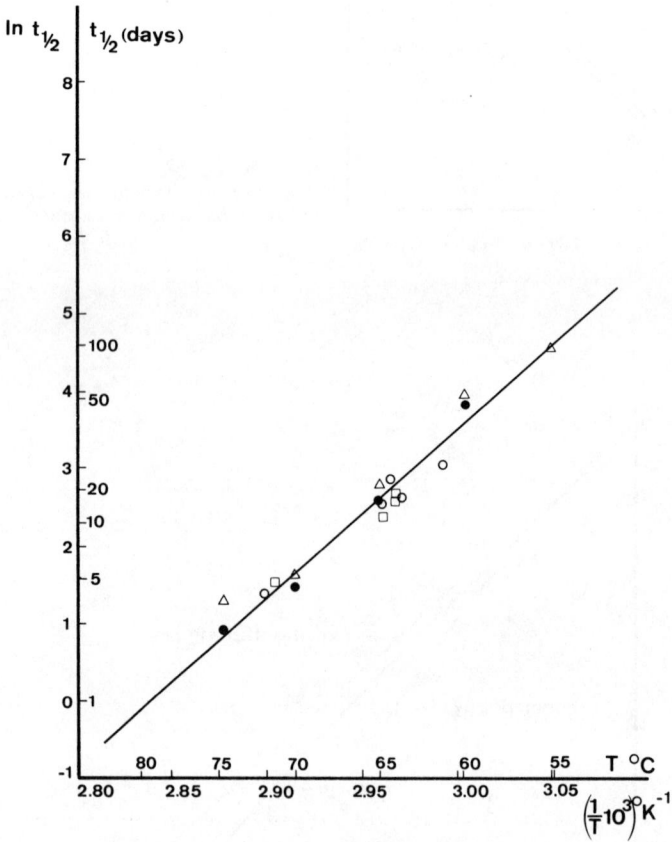

Figure 9. *Experimental data and theoretical relationship between fixed-bed activity. (●) Constant conversion, (△, □, ○) constant flow.*

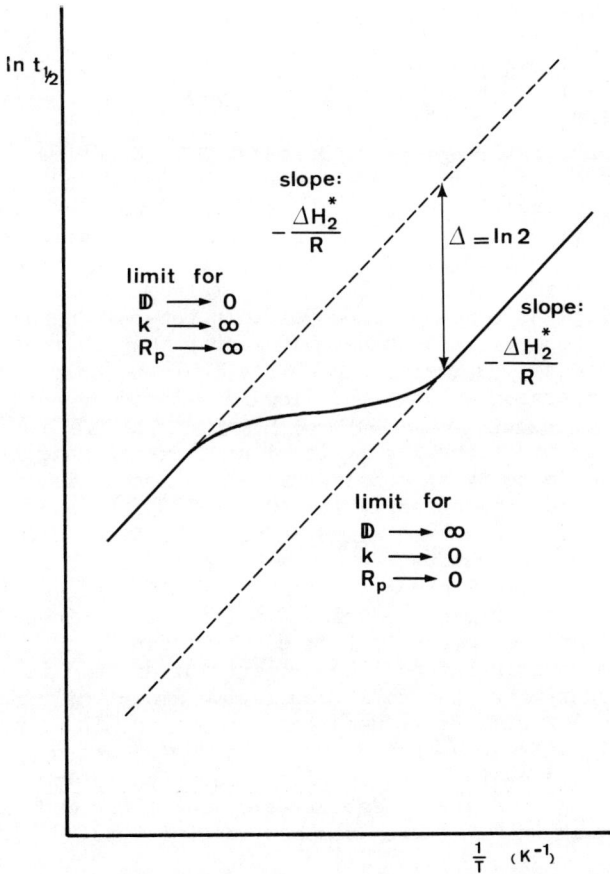

Figure 10. Overall characteristics of the Arrhenius plot of initial flow velocity half life

properties of the particles and in particular the particle radius
and the glucose diffusivity in the particles. The significance of
diffusional limitation to the half life of the immobilized enzyme
used in the present investigation is illustrated in figure 11
where the ratio of the activity half life of the immobilized enzyme to the estimated activity half life of the free enzyme is
shown as a function of temperature. The immobilized enzyme activity half life is higher by 10 to 60% than the half life of the free
enzyme activity. One of the implications of the remarks made above
is that estimation of the activation enthalpy of the deactivation
of the free enzyme from an Arrhenius plot of the flow velocity
half life of the immobilized enzyme in a fixed bed is, in general,
not justified.

One remark still needs to be made. Most experiments were performed at a constant flow rate through the fixed bed and the decrease in the fraction of glucose isomerized was observed as a
function of time. In the foregoing discussion we referred to the
half life of the initial flow velocity at constant relative conversion and this is definitely different from the relative conversion half life at constant flow rate. In general the half life of
the latter property is considerably longer than that of the former.
The present model furthermore predicts that the half life of the
relative conversion at constant flow rate is dependent on the initial glucose concentration and the initial relative conversion. On
inspection of equation (17) it is clear that the half life of the
initial flow velocity at constant relative conversion is equal to
the half life of the property $\ln ((C_{S0} - C_S^x) / (C_{SE} - C_S^x))$ at constant flow rate and our calculations of the initial flow velocity
half life from the constant flow experiments were based on this
equivalency. Figure 9 shows that the half life estimates from the
constant flow experiments obtained in this way do indeed agree
with those obtained directly from constant conversion experiments.

A commercially interesting property of an immobilized enzyme
is its productivity, the total cumulative amount of syrup converted, during the time it is used for production. In principle the
relationship between amount of syrup converted and operating time
can be obtained by integration of the equation for the flow velocity resulting in a given percentage of relative conversion of
glucose (equation (17)) with respect to time i.e.:

$$P(t) = \int_0^t V(t)\, dt \qquad\qquad (19)$$

in which $V(t)$ is given by equation (17), the values of K is calculated as a function of operating time from equation (13) and the
efficiency factor is calculated from equation (8).

The Thiele factor to be substituted in equation (8) is given
as a function of time by the following equation:

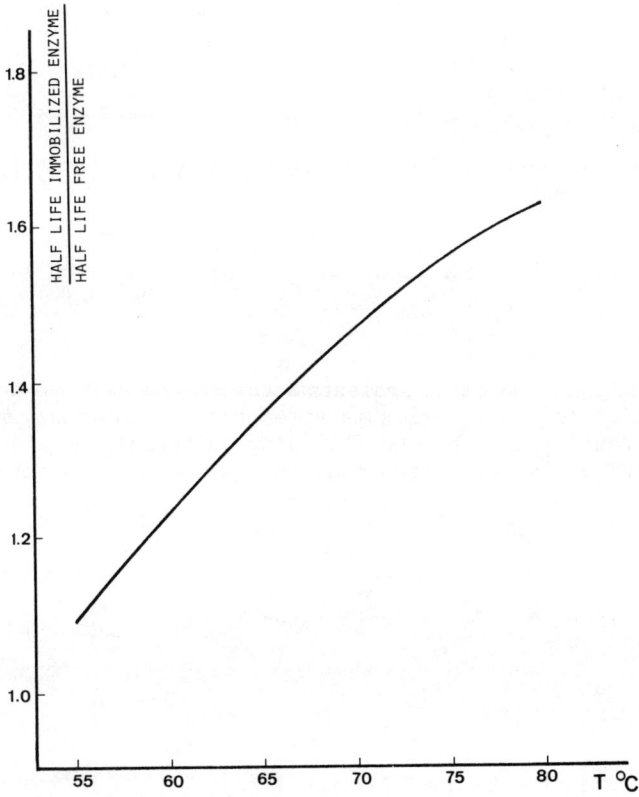

Figure 11.   Theoretical relationship on ratio between activity half life of free and immobilized enzyme

$$\Phi = \Phi_o\, e^{-\frac{1}{2}\, k_d t} \tag{20}$$

in which Φ_o is the initial Thiele factor being given by:

$$\Phi_o = R_p \sqrt{\frac{K_o}{D}} \tag{21}$$

By combination of equations (21), (20) and (13) it can be shown that the momentary value of K can also be written as:

$$K = K_o \frac{\Phi^2}{\Phi_o^2} \tag{22}$$

Equation (19) can now with the aid of equations (17), (21) and (22) be evaluated to:

$$P(t) = \int_0^t \frac{D(1 - \varepsilon)\,.\,A.H}{R_p^2 \ln \dfrac{C_{sO} - C_s^x}{C_{sE} - C_s^x}} \,.\, 3\Phi\,(\frac{1}{\tanh\Phi} - \frac{1}{\Phi})\, dt \tag{23}$$

Equation (23) still presents a rather complex problem if straightforward integration is attempted. It can however be simplified considerably by the following manipulations
First it is recognized that the following relationship holds:

$$dt = \frac{dt}{d\Phi}\,.\,d\Phi \tag{24}$$

From equation (20) $\dfrac{dt}{d\Phi}$ is calculated to be

$$\frac{dt}{d\Phi} = \frac{-2}{k_d \Phi} \tag{25}$$

With the aid of (25) and (24) the integral (23) is transformed to:

$$p(\Phi) = \int_{\Phi_o}^{\Phi} - \frac{6D\,(1 - \varepsilon)\,.\,A.H}{k_d R_p^2}\,(\frac{1}{\tanh\Phi} - \frac{1}{\Phi})\, d\Phi \tag{26}$$

in which $p(\Phi)$ is the productivity at the time t the Thiele factor
has decreased to Φ $(m^3/syrup)$
Performing the integration in (26) results in the following expression:

$$p(\Phi) = \frac{6D(1 - \epsilon) \cdot A.H}{k_d R_p^{2}} \left\{ \ln\left[\frac{\sinh\Phi_o}{\Phi_o}\right] - \ln\left[\frac{\sinh\Phi}{\Phi}\right] \right\} \quad (27)$$

$p(\Phi)$ gives the productivity of a column of A.H m^3; a more useful property is obtained if $p(\Phi)$ is expressed per unit of column volume:

$$P(\Phi) = \frac{6 D(1 - \epsilon)}{k_d R_p^{2}} \left\{ \ln\left[\frac{\sinh\Phi_o}{\Phi_o}\right] - \ln\left[\frac{\sinh\Phi}{\Phi}\right] \right\} \quad (28)$$

in which $P(\Phi)$ is the productivity at the time t the Thiele factor has decreased to Φ (m^3 syrup/m^3 column)

With the aid of the equation for the relationship between Thiele factor and time, equation (20), equation (28) can now be used to calculate productivity as a function of time at various operating temperatures. (In evaluating (28) it has to be borne in mind that k_d, which was expressed in 1/day, has to be transformed to 1/s before substitution in equation (28)).

The results of the evaluation of (28) for an immobilized glucoseisomerase defined by the properties given in table I have been given in figure 12. The productivity refers to a situation in which a 45% w/w glucose syrup is isomerized to HFCS with a fractional conversion of glucose of 45%. Some of the experimental results obtained by integration of the flow velocity time curves of the constant conversion experiments are shown in figure 12.

Figure 12 shows that if one has all the time in the world and the inclination to wait endlessly productivity will eventually be highest at the lowest temperature in the range studied. At 50°C productivity is about 7 times higher than that at 65°C. If one decides to use the enzyme only for 100 days maximal productivity is obtained at 55°C, it is twice that at 65°C. If patience runs out after 50 days the maximum productivity is obtained at 60°C, productivity exceeds that at 65°C by 30%.

It will be clear that in actual practice, the optimal operating temperature depends, apart from problems of microbial contamination which becomes increasingly significant at temperatures below 60°C, on the relative costs of the enzyme preparation per unit of column volume and the operating costs per unit volume. Another important limitation may be the total amount of HFCS to be produced with a given capacity in terms of column volume, this will to a large extent be determined by the demand for HFCS. At times of low demand the optimal production temperature will tend to be lower than in times of high demand.

In figure 13 it is qualitatively shown how the optimal operating temperature and operating time can be estimated once the

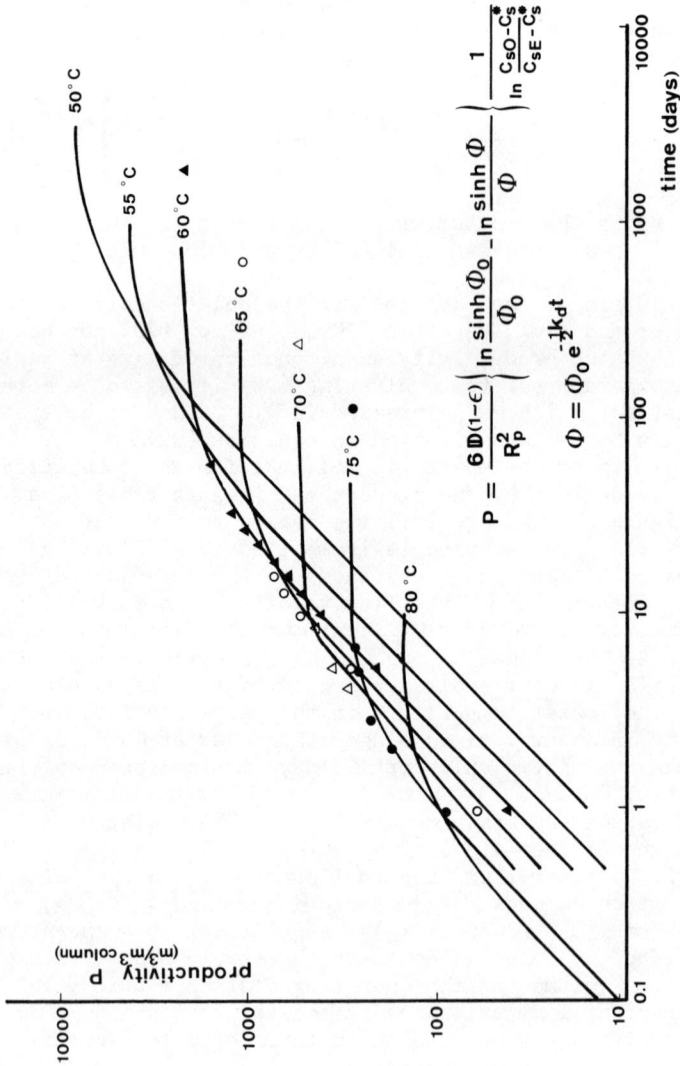

Figure 12. Productivity vs. operating time at various temperatures

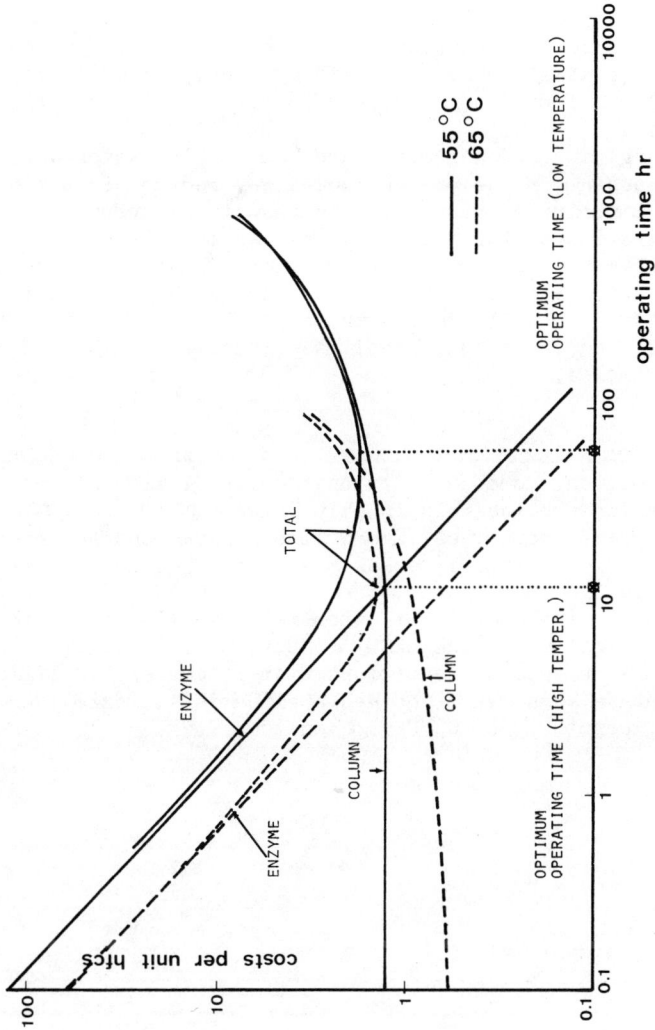

Figure 13. Qualitative plot of contributions to HFCS production costs at two operating temperatures: (———) *55°C;* (— — —) *65°C.*

costs of column operation and the costs of the immobilized enzyme are known. It will be clear that once realistic values of the cost contributions are known the present model allows the optimization of HFCS production costs.

One qualitative conclusion the model permits is that, in principle, economically more favourable production of HFCS is possible if an increasing instead of a constant temperature with operating time is applied.

The main problems associated with the use of the present model for large scale operation are that it was tested using laboratory scale columns and a "syrup" consisting of a solution of pure dextrose.

At a large scale, channeling and back mixing which were assumed to be absent in the present model may reduce productivity. Furthermore impurities present in the industrial syrups may affect the kinetics of the various processes involved. Another serious problem may be plastic deformation of the particles under the pressure necessary to force the syrup through the fixed bed, which may eventually reduce the area for mass transfer to the particles and cause additional mass transfer resistances outside the particles.

Abstract

A mathematical model permitting the formulation of an analytical expression for the relationship between production rate of HFCS, free enzyme properties, immobilized enzyme characteristics, temperature and operating time has been developed and verified experimentally.

The theory developed is of potential significance to the economic optimization of the HFCS production process. It is shown that under most practical conditions Gist-Brocades' immobilized glucoseisomerase, has, in fixed bed applications, maximal productivity at temperatures of $60^{\circ}C$ or even lower.

List of symbols

A_1, A_2	constants	$(1/s)$
A	column cross section	(m^2)
c_E	the concentration of organism dry matter	$(kg\ DM/m^3)$
c_s	glucose concentration	$(mole/m^3)$
c_s^{*}	glucose concentration at thermodynamic equilibrium	$(mole/m^3)$
c_{si}	glucose concentration at particle interface	$(mole/m^3)$
c_{s0}	glucose concentration at column inlet	$(mole/m^3)$
c_{sE}	glucose concentration at column exit	$(mole/m^3)$
D	diffusion coefficient for glucose in particles	(m^2/s)
$[E]$	intrinsic enzyme concentration	$(mole/kg\ DM)$
h	height coordinate	(m)
H	column height	(m)
ΔH_1^{*}	activation enthalpy of conversion of glucose to fructose	$(kJ/mole)$
ΔH_2^{*}	activation enthalpy of enzyme deactivation	$(kJ/mole)$
k_1, k_2, k_{-1}, k_{-2}	rate constants	
k	first order rate constant	$(1/s)$
k_d	first order rate constant for deactivation of free enzyme activity	$(1/day)$
K	pseudo first order rate constant	$(1/s)$
K_o	initial value of pseudo first order rate constant	$(1/s)$
$K_{0,65}$	initial value of pseudo first order rate constant at $65^{\circ}C$	$(1/s)$
K^{*}	equilibrium constant	$(-)$
K_s'	pseudo Michaelis-Menten constant	$(mole/m^3)$
K_{MG}	the Michaelis-Menten constant for the conversion of glucose to fructose	$(mole/m^3)$
K_{MF}	the Michaelis-Menten constant for the conversion of fructose to glucose	$(mole/m^3)$
$p(t)$	cumulative amount of syrup converted at time t	(m^3)
$p(\Phi)$	cumulative amount of syrup converted at time corresponding to a Thiele factor of (Φ)	(m^3)
$P(\Phi)$	total amount of glucose solution converted per unit enzyme at time corresponding to a Thiele factor (Φ)	$(m^3/m^3\ enzyme)$
r_s	glucose conversion rate	$(mole/m^3 s)$
$r_{s,\ max}'$	the apparent maximal forward rate of reaction of glucose to fructose	$(mole/m^3 s)$
R	universal gas constant	$(kJ/mol^{\circ}K)$
R_p	radius of immobilized enzyme particles	(m)

t	time	(m)
$t_{\frac{1}{2}}$	activity half life of the immobilized enzyme	(day)
T	absolute temperature	(^{O}K)
V	volume of immobilized enzyme	(m^3)
V^e_o	initial volumetric flow rate resulting in 45% conversion of glucose	(m^3/s)
$V_{0,65}$	initial volumetric flow rate resulting in 45% conversion of glucose at 65°C	(m^3/s)
$V'_{s, max}$	the maximum specific rate of fructose formation	(mole/kg DM s)
ε	packed bed porosity	(-)
η	efficiency factor	(-)
Φ	Thiele factor	(-)
Φ_o	initial Thiele factor	(-)

Literature cited

1. Michaelis, L. and Menten, M.L., Biochem. Z. (1913) 49, 333.
2. Heineken, F.G., Tsuchiya, H.M, and Aris, R., Math. Biosci. (1967) 1, 95.
3. Fratzke, A.R., Lee, Y.Y. and Tsao, G.T., G.V.C./AICHE-Joint Meeting, München (1974) 4, F2-1.
4. Satterfield, C.N., "Mass Transfer in heterogeneous catalysis". Cambridge University Press, Cambridge (1970).
5. Vellenga, K., "The isomerization of D-glucose into D-fructose" Ph. D. Thesis (1978) State University of Groningen. The Netherlands.
6. Nielsen, M.H.,"Proceedings of Chemical Engineering in a changing world". (1976) 183. Elsevier, Amsterdam.

RECEIVED February 15, 1979.

Evaluation of a Novel Microporous PVC–Silica Support for Immobilized Enzymes

Its Use in a Flow-Through Reactor System for Production of Fructose

BRUCE S. GOLDBERG, ALEXANDER G. HAUSSER, KEVIN R. GILMAN, and RICHARD Y. CHEN

Amerace Corporation, Microporous Products Division, Ace Road, Butler, NJ 07405

During the past fifteen years there has been a great deal of research in the area of immobilized enzymes. This research effort has been focused in three major areas. First is the form of the enzyme to be immobilized, whether the enzyme is purified or is contained within whole cells. The second has been in the area of immobilization techniques, whether by physical absorption, entrapment, or chemical bonding. The third area of research has been in the development of various types of carriers that can support the enzyme and be used in various types of reactors.

The purpose of this paper is to discuss the third area, viz. the enzyme support. Various carriers that have been used over the years for immobilizing enzymes can be classified into three categories. The first is hard particulate substances such as porous glass/ceramics and polymers. The second category is polymers in membranous form, such as reconstituted collagen or ultrafiltration membranes, where the enzyme is trapped behind or within the membrane barrier. The third category is cellulose-derived materials in the form of fibers or beads. Almost all these materials are used either in the form of packed beds or as membranes. In any case, the diffusional resistances are major restrictions to their use as efficient enzyme supports. We will discuss and demonstrate a new type of microporous carrier that can be used very efficiently as an immobilized enzyme support.

Support Characteristics

Amerace Corporation has developed a novel microporous support which overcomes the disadvantages of supports used previously. This material can be used in a flow-through reactor in which the substrate flow is perpendicular to the surface of the support; ie, the substrate passes through the support material, rather than around it. Reaction takes place as a result of the

substrate's directly contacting the enzyme through bulk flow,
rather than its having to diffuse into the support before it can
contact the enzyme.

Briefly, the flow-through reactor has the advantages of
essentially no external or internal mass transfer limitations, no
substrate holdup, and no channeling; all of which contribute to a
more efficient utilization of immobilized enzyme. The substrate
must contact all reactive sites while passing through the pores;
therefore eliminating dispersion or diffusion related problems.
One can perform sequential reactions, since there is no holdup of
reactants and products from one reaction step to the next.

The support is a microporous PVC-silica sheet having a porosity in the 70-80% range. The pore size as determined by mercury
intrusion porosimetry is in the 0.2 μm to 2.0 μm range. The
support is extremely hydrophilic, has a negative charge, and a
surface area of 80 m^2/g. The material is non-compressible under
normal conditions, is steam sterilizable, and has a low dry
density of 0.45 g/cm^3. The microporous support has received FDA
approval for direct food contact. The tortuosity of the pore
structure requires that the substrate make intimate contact with
the active enzyme as it passes through the support material. The
active sites are attributed to the silica contained within the
porous matrix which allows the addition of organic functionality.

While this paper deals with the chemistry and reaction
characteristics of glucose isomerase, it should be noted that
glucose oxidase, glucoamylase, aldose-1-epimerase, α-glucosidase,
alcohol dehydrogenase, pullulanase, and fungal α-amylase have
been successfully immobilized onto this support.

Enzyme Bonding Procedure

The support preparation and immobilization procedures are as
follows. Initially, the sheet is activated by soaking for one
hour in a 1 M NaCl solution containing 1.5% polyethyleneimine
(PEI) having a molecular weight of approximately 40,000. The
support material is rinsed free of excess PEI and may be dried
for later use. The support material may be cut to the proper
size and configuration before or after treatment with polyethyleneimine. All work reported in this paper, unless otherwise
stated, was done with 47 mm diameter discs having a thickness of
approximately 0.5 mm. The discs were mounted in reactors comprised of 47 mm Millipore Swinnex filter housings or were molded
into a stacked disc configuration, the concept of which is pre-
sently the subject of a patent application. The reactor system
is assembled as illustrated in Figure 1. The system is first
filled with distilled water and a 2.5% solution of glutaraldehyde,
at pH 9.5, is pumped through the reactor at 6 ml/min. for 1 hr.
Excess glutaraldehyde is removed from the reactor by flushing with
pH 7.5 buffer solution containing 2 g/l $MgSO_4 \cdot 7H_2O$ and 2 g/l
$NaHCO_3$. Buffer is pumped through the reactor until the effluent

has maintained a pH of 7.5 for 15 minutes. When the effluent pH has been stabilized at 7.5 the enzyme solution is recycled through the reactor for 1 hr. Excess enzyme is recovered by flushing the reactor with the above buffer and collecting the effluent. This entire procedure is carried out at room temperature.

In the above sequence, the PEI is chemi-adsorbed onto the surface of the silica particles in the PVC matrix and confers organic functionality to it. The glutaraldehyde crosslinks the PEI rendering it totally insoluble and also acts as a leash for the enzyme pendent group with which it is subsequently reacted.

The enzyme, which in this case is glucose isomerase, was obtained from Novo Laboratories and is derived from <u>Bacillus coagulans</u> NRL-5666. The soluble enzyme was isolated from a dried cell preparation, as supplied by Novo Laboratories, by suspension in buffer and centrifugation to remove cellular material and other insoluble impurities. The resulting solution was fractionated by ammonium sulfate precipitation retaining the 55-70% fractions followed by dialysis to obtain a purified enzyme suitable for immobilization onto the microporous support.

The purity of the above prepared enzyme is approximately 7,000 IGIU/g. One IGIU is defined as the amount of enzyme required to convert one micromole of glucose to fructose in one minute at 60°C in a solution containing 40% w/w glucose, 2 g/l $MgSO_4 \cdot 7H_2O$, 1 g/l $NaHCO_3$ having a pH of 7.5 (measured at room temperature). The activity of soluble enzyme solutions were measured on a Yellow Springs biological oxygen analyzer (Model 53) which measures the enzyme catalyzed oxidation of glucose to gluconic acid and by liquid chromatography of glucose-soluble enzyme reaction products.

Depending on the purity and the history of the dry enzyme preparation, recovery yields of 90-100% expressed activity of the immobilized enzyme have been achieved. The activity of the soluble enzyme preparation after dialysis is approximately 400 IGIU/ml and a 2X quantity of enzyme is utilized for immobilization with roughly 50% of the offered soluble enzyme being recovered after immobilization and 45-50% of the offered soluble enzyme resulting in expressed activity on the reactor when operated under the above conditions. Depending upon the activity of the purified enzyme, the activity of the carrier after immobilization is typically 600 IGIU/g.

Analysis of Diffusional Limitations

As mentioned earlier, one of the important features of this immobilized enzyme support is the fact that it permits design of a flow-through reactor where the substrate must pass through all the pores. As a result, the diffusion limitations of the support are reduced to a minimal level. An experiment was performed to determine the minimum velocity below which diffusional effects

become evident. All subsequent experiments were run at velocities above the minimum value.

Figure 1 illustrates the experimental setup used to determine this critical velocity. A 47 mm diameter reactor, having an
expressed activity of 145 IGIU/g was placed in a 60°C temperature
controlled bath. Substrate was recirculated through this reactor
at various velocities and the time required for the 108 ml of
reactor and reservoir volume to reach 5% conversion at each
superficial velocity was recorded and the expressed activity
calculated. Figure 2 illustrates a plot of the resulting data.
Typically, above a superficial velocity of 0.1 cm/min the expressed activity remained constant at 145 IGIU/g.

Kinetic Evaluation

To illustrate the efficiency of enzyme utilization with this
microporous enzyme support, 30 g of support material was immobilized with 22,000 IGIU of purified enzyme. The resulting
reactor had an expressed activity of 726 IGIU/g at 60°C. A 40%
w/w solution of 99% pure dextrose containing 2 g/l $MgSO_4 \cdot 7H_2O$ and
1 g/l of $NaHCO_3$ at pH 7.5 was converted to a 45% fructose product
in 6.5 min and an equilibrium product of approximately 50% fructose in less than 12 min residence time. The flow rate required
to obtain a 45% converted product was 6 ml/min.

A second experiment was performed to demonstrate the
superior utilization of enzyme by a microporous sheet flow-
through reactor versus a packed column containing controlled pore
glass. Controlled pore glass particles 40-80 mesh were obtained
from Electro Nucleonics Corporation and chemically modified by
standard methods[1] to introduce covalently bound, aliphatic
amino functionality on the external and internal surfaces. Two
grams of the amino-CPG were degassed and suspended in 100 ml of
10% pH 8 glutaraldehyde solution for 1 hr. The CPG particles
were extensively washed to remove excess glutaraldehyde using
pH 7.5 Hepes buffer containing 2 g/l $MgSO_4 \cdot 7H_2O$. Ten ml of
enzyme solution containing 0.43 units/ml at pH 7.5 was added to
the particles and allowed to react for 1 hr with gentle agita-
tion. Excess enzyme was rinsed from the CPG particles and based
upon loss of activity from the enzyme solution, the loading was
0.66 units/ml (CPG has a bulk density of 0.36 g/ml). The 1.2 ml
volume of the particles was loaded into a small 0.6 cm diameter
column and used in the subsequent experiment.

Four 26 mm diameter MPS discs were treated with PEI, mounted
in a standard 26 mm Millipore Swinnex filter housing and set up
as previously described. A 10% pH 8 solution of glutaraldehyde
solution was pumped through the reactor for 1 hr. The reactor
was rinsed with 200 ml of pH 7.5 Hepes buffer containing 2 g/l
$MgSO_4 \cdot 7H_2O$ for 1 hr. Enzyme solution, 30 ml having an activity
of 0.43 units/ml, was recirculated through the reactor for 1 hr
after which time excess enzyme was flushed from the reactor using

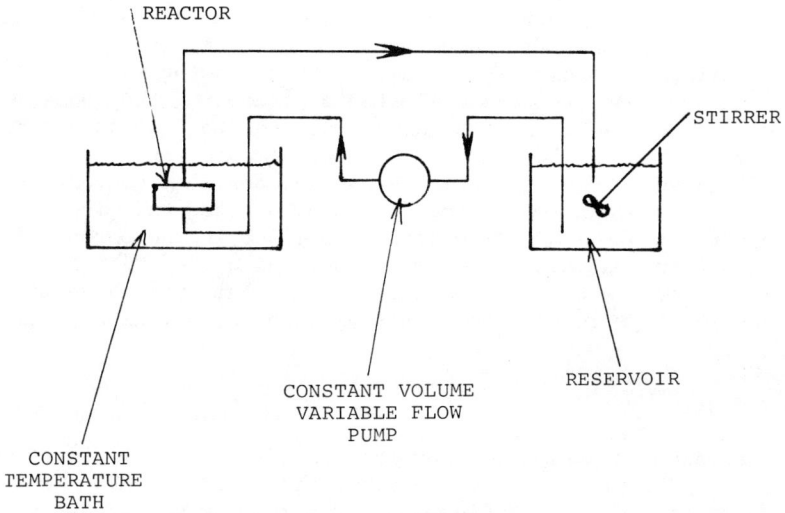

Figure 1.   *Flow diagram of reactor system for diffusion studies*

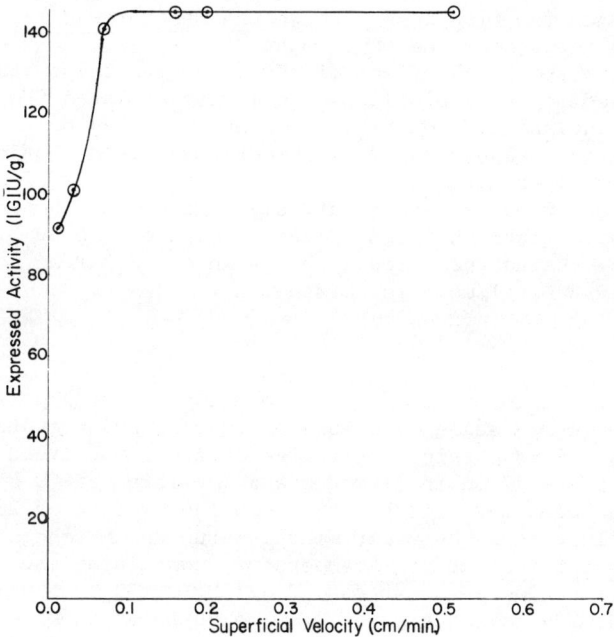

Figure 2.   *Effect of superficial velocity on enzyme utilization efficiency. Reactor size: diameter—47 mm, thickness—0.45 mm, weight—0.43 g. Conditions: pH— 7.5, temperature—60°C. Substrate: dextrose—45% w/v, $MgSO_4 \cdot 7H_2O$—2 g/L, $NaHCO_3$—1 g/L.*

the above buffer solution. The loading on the reactor was calculated to be 0.7 units/ml based upon loss of activity from the
immobilizing solution. The volume of this reactor was 1.0 ml.

The two reactors were evaluated by measuring the degree of
conversion of a 7.2% w/v fructose substrate (ph 7.0) at several
flow rates when operated in a single pass mode. The resulting
data is shown in Figure 3 which shows the per cent conversion
obtained versus residence time. It can be noted that it takes
5.5 min to obtain a 19.1% conversion product in the porous glass
packed column, whereas the reactor containing the MPS immobilized
enzyme achieves a conversion of 34% in the same residence time.
The MPS is 78% more efficient in utilizing an equivalent activity
of enzyme.

pH Study

The experimental setup for the evaluation of the effect of
pH and temperature on the expressed activity of the MPS immobilized enzyme is shown in Figure 4. The reactor was 47 mm in
diameter and contained 2 g of support material having an expressed activity of approximately 650 IGIU/g. An approximate 10%
converted product was obtained at 60°C and pH 7.5 on a substrate
containing 40% w/w dextrose, 2 g/l $MgSO_4 \cdot 7H_2O$, 1 g/l $NaHCO_3$, and
1 g/l $NaHSO_3$. The flow rate through the reactor was maintained
at 5 ml/min throughout the experiment.

Figure 5 shows the effect of pH at varying temperatures on
the expressed activity of glucose isomerase isolated from
Bacillus coagulans and immobilized on MPS. It may be noted that
the optimum pH (measured at room temperature) is not affected by
varying the temperature and that the per cent change of activity
as a function of pH remains relatively constant with changing
temperatures. Figure 6 is an Arrhenius plot of the pH 7.5 data
in Figure 5. Using the formula $\ln K = \ln A - E_a/RT$ an activation
energy of 19.6 K cal/mole is obtained.

Half-Life Study

Using the information from the pH-temperature experiment we
then built a large scale flow-through reactor and operated it at
60°C and pH 7.5 to obtain a half-life on the immobilized enzyme.
This reactor was 47 mm in diameter and consisted of 80 stacked
sheets of support material having a combined weight of 30 g.
Figure 4 illustrates the experimental setup which consisted of a
reservoir maintained at room temperature containing the substrate
which contained 40% w/w dextrose, 2 g/l $MgSO_4 \cdot 7H_2O$, 1 g/l $NaHCO_3$,
and 1 g/l $NaHSO_3$ having a pH of 7.5 (measured at room temperature). This substrate was pumped by a variable speed pump
through a prefilter containing an 8 µm paper filter and an
Amerace 0.45 µm filter and then through the reactor which was
placed in a 60°C constant temperature water bath. The substrate

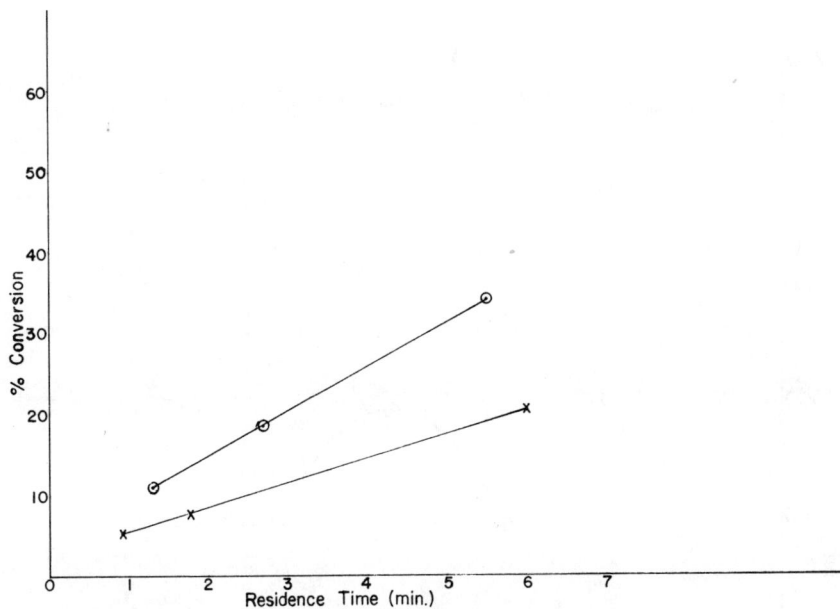

Figure 3.   *Percent conversion vs. residence time. Substrate: fructose—7.2% w/v (4M), pH—7.5, temperature—25°C. (○) MPS flow-through reactor, (×) controlled pore-glass reactor.*

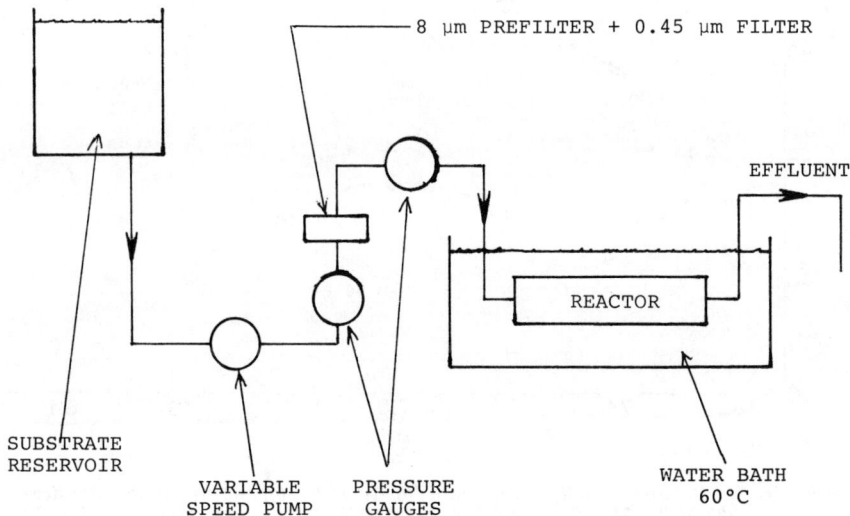

Figure 4.   *Flow diagram of reactor system for constant conversion of dextrose*

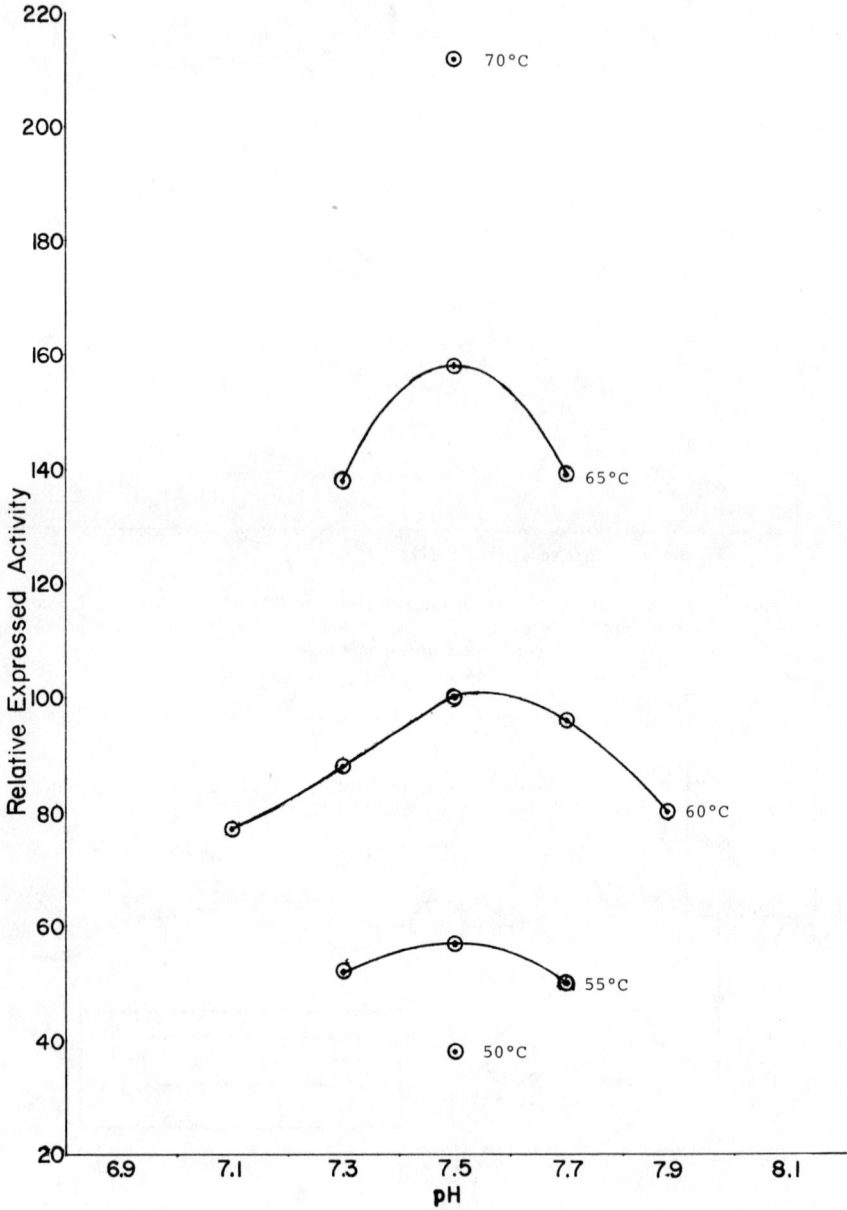

Figure 5.   Relative activity vs. pH as a function of temperature. Substrate: dextrose—40% w/w, MgSO₄ · 7H₂O—2 g/L, NaHCO₃—1 g/L, NaHSO₃—1 g/L.

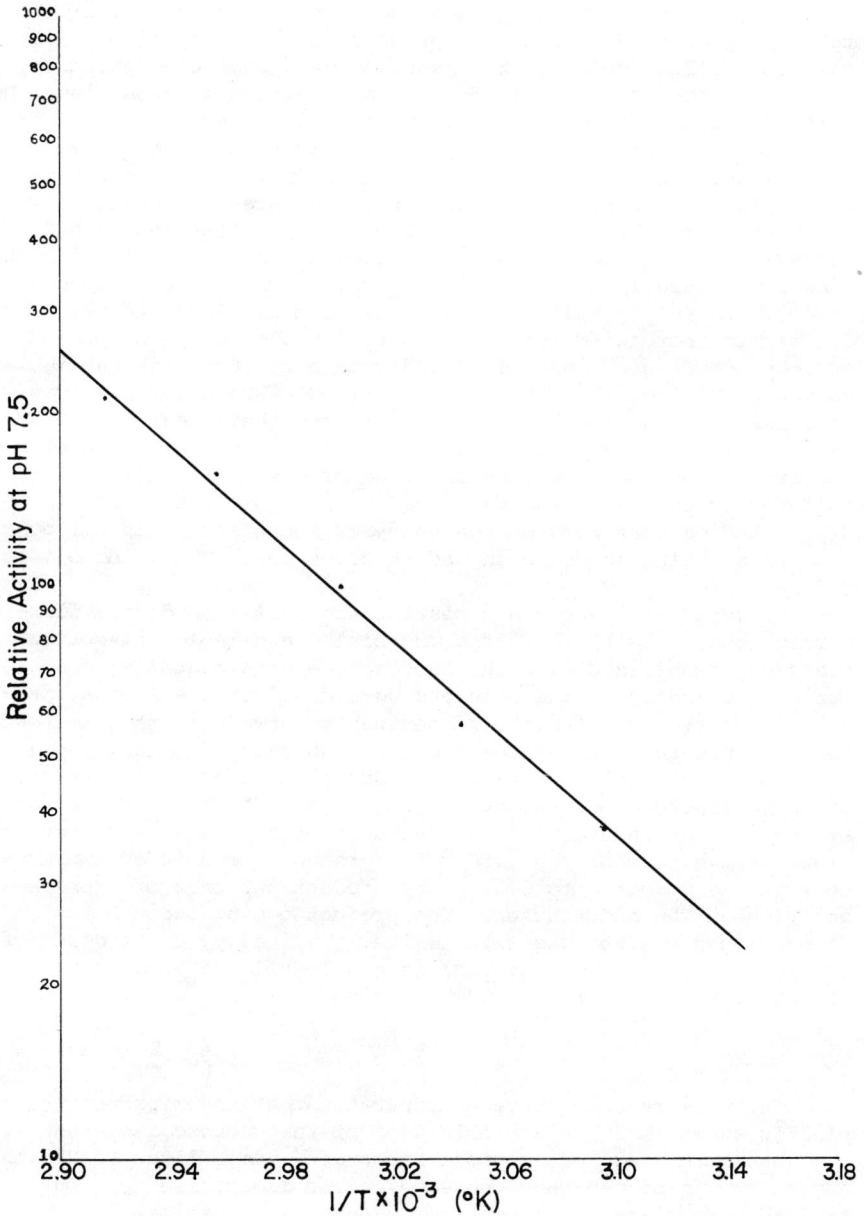

Figure 6. Arrhenius plot for Bacillus coagulans in flow-through reactor. Substrate: dextrose—40% w/w, $MgSO_4 \cdot 7H_2O$—2 g/L, $NaHCO_3$—1 g/L, $NaHSO_3$ —1 g/L, pH—7.5.

passing through the reactor was discharged into a reservoir and
analyzed daily. Pressure gauges were located before and after
the prefilter to determine the pressure or contaminant build-up
on the prefilter or reactor. As is normal commercial practice,
the conversion was maintained at a constant 45% fructose level by
suitably adjusting the flow rate through the reactor.

Figure 7 shows the experimental activity decay curve for
this immobilized enzyme reactor system. Initial residence time
to obtain a 45% converted product from the starting dextrose
substrate was in the order of 10.6 min. The first and second
half-lives ($t_{1/2}$ and $t_{1/4}$) were defined as the times at which the
flow rates were 1/2 and 1/4 of the initial flow rate, respectively. In this particular case, the initial flow rate was
5.1 ml/min increasing to about 5.6 ml/min over a 200 hr period
and then decreasing in a first order manner. For this particular
enzyme at 60°C on this support $t_{1/2}$ was 920 hrs while $t_{1/4}$ was
1600 hrs. During the half-life experiment there was no problem
with pressure buildup either on the prefilter or on the reactor
and the effluent coming from the reactor was water white. This
buffered system did not show any pH drop across the reactor.
Other studies performed on non-buffered substrates with all other
conditions being the same showed pH drops in the range of 0.5 to
0.8 pH units.

An experiment was run comparing the half-life of the flow-
through reactor with that of commercially available polymeric
granules immobilized with the same enzyme strain utilizing a
packed bed reactor. The reactors were operated the same as previously described. Table I summarizes the two systems. We found
that the productivity of the flow-through reactor calculated at
$t_{1/4}$ was in the order of 4100 g of dry product/g of support compared to 1500 g of dry product/g of support for the bed reactor
system. On an enzyme utilization basis, the productivity for the
flow-through reactor was 8.56 g dry product per unit of expressed
activity as compared to 6.54 g dry product per unit of expressed
activity on the bed reactor. The product coming through the
flow-through reactor was water white. Typically, the product of
the packed bed reactor material is highly colored and has a
strong odor for the first 100 hrs.

Discussion

The work reported here demonstrates that the microporous
plastic sheet (MPS) is a viable support for glucose isomerase.
Bacillus coagulans glucose isomerase immobilized on MPS has
a greater efficiency than the same enzyme immobilized on controlled pore glass. The controlled pore glass exhibits an
apparent efficiency of 0.56 relative to MPS. Pore diffusion
limitations of the glass beads and bed channeling cause this
inefficiency.
The ratio of the productivity of Bacillus coagulans glucose

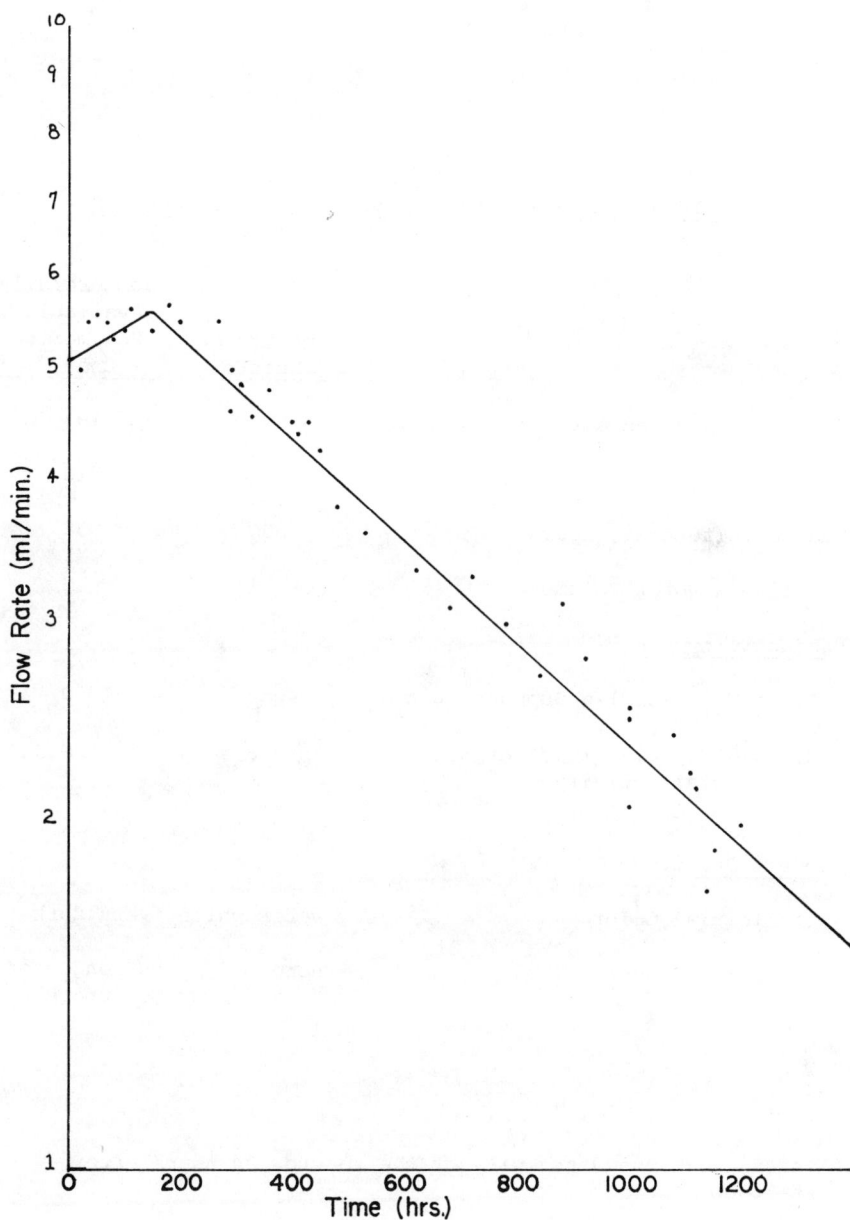

Figure 7.   Half life study of glucose isomerase on Amerace flow-through reactor. Conditions: pH—7.5, temperature—60°C.   Substrate: dextrose—40% w/w, $MgSO_4 \cdot 7H_2O$—2 g/L, $NaHSO_3$—1 g/L, $NaHCO_3$—1 g/L.

Table I

COMPARISON OF FLOW-THROUGH VERSUS PACKED BED REACTOR

	Amerace MPS Reactor	Commercially Available Polymeric Beads
Initial Expressed Activity, IGIU*/g	572	193
Half-Life $t_{1/2}$, hours	920	710
Quarter-Life $t_{1/4}$, hours	1600	938
Relative Reactor Volume	1	2
Productivity		
1. g HFCS (dry)/g Support at $t_{1/4}$	4064	1450
2. g HFCS (dry)/IGIU* at $t_{1/4}$ Initial Activity	8.56	6.54
Product Quality		
1. Initial Color	water white	very dark
2. Initial Odor	none	strong odor of protein

* Initial expressed activity of the reactor 24 hours after start up

isomerase immobilized in commercially available polymeric beads
to that enzyme immobilized on MPS is 0.76. The lower productivity of the polymeric beads is apparently due to a combination of
pore diffusion limitation, other factors inherent in the nature
of the polymeric beads, and the method by which they are
utilized.

Since pore diffusion is limited and channeling is at a
minimum, MPS allows for more efficient utilization of enzyme.
Shorter reaction times are possible with MPS due to the use of
the flow-through reactor concept and the higher activities
attainable.

Summary

From a reactor engineering-design standpoint the flow-
through reactor would require about one-half the size of a packed
bed reactor. This ratio can be even greater depending on the
specific activity of the enzyme utilized in the immobilization.
The higher the enzyme purity, the smaller the size of the flow-
through reactor. Other advantages of the flow-through reactor
are rapid and efficient change-over of reactors and the elimination of compressibility and other hydraulic problems usually
associated with packed bed reactors.

This development is the subject of a recently issued U. S.
patent[2] and other pending patent applications.

ABSTRACT

EVALUATION OF A NOVEL MICROPOROUS PVC-SILICA SUPPORT FOR IMMOB-
ILIZED ENZYMES AND ITS USE IN A FLOW-THROUGH REACTOR SYSTEM FOR
PRODUCTION OF FRUCTOSE

A microporous polyvinyl chloride-silica filled plastic sheet
was evaluated as an immobilized enzyme support using a flow-
through reactor concept. The substrate was pumped through the
microporous plastic sheet (MPS) containing immobilized glucose
isomerase chemically bound to the internal pore structure. The
physical properties of the plastic sheet and the advantages of
its use as a flow-through reactor as compared to a packed bed
reactor are presented. The immobilization chemistry is described
along with a listing of various enzymes which have been successfully immobilized onto the support. A detailed study, including
half-life, is reported on the production of fructose utilizing
glucose isomerase isolated from Bacillus coagulans. Reaction
kinetics including the effects of temperature and pH on the
relative activity of the immobilized enzyme was studied and the
activation energy calculated. A minimum velocity was determined
above which the diffusional affects in the microporous sheet are

eliminated. Typical operating conditions for the immobilized
flow-through reactor were 60°C, a pH of 7.5 (at 25°C), 40% w/w
dextrose substrate containing 1 g/l of $NaHCO_3$, 1 g/l $NaHSO_3$,
2 g/l $MgSO_4 \cdot 7H_2O$.

Literature Cited

(1) Weibel, M. K. et al, "Immobilized Enzymes: A Prototype Device
for the Analysis of Glucose in Biological Fluids Employing
Immobilized Glucose Oxidase," Anal. Biochem., 52 502 (1973)

(2) Goldberg, Bruce S. (assigned to Amerace Corporation)
"Immobilized Proteins," U. S. Patent #4,102,746, July 25,
1978

RECEIVED March 29, 1979.

Immobilized Microbial Cells with Polyacrylamide Gel and Carrageenan and Their Industrial Applications

ICHIRO CHIBATA

Research Laboratory of Applied Biochemistry, Tanabe Seiyaku Co., Ltd., 16-89, Kashima-3-chome, Yodogawa-ku, Osaka, Japan

Enzymes are very efficient and advantageous catalysts, and can catalyze specific reactions under mild conditions, in neutral aqueous solution at room temperature. However, they are not always ideal catalysts for industrial application. In some cases, the above mentioned advantages of enzymes turn to be disadvantageous as catalysts. Namely, enzymes are generally unstable, and can not be used in organic solvents and at elevated temperatures.

One of the approaches to prepare more superior catalysts for application purpose is immobilization of enzymes. Over the past 10 years, the immobilization of enzyme has been the subject of increased interest, and many papers on potential applications of immobilized enzymes and microbial cells have been published. However, practical industrial systems using immobilized enzymes and immobilized microbial cells have been very limited.

In 1969, we [1, 2] succeeded in the industrial application of immobilized enzyme, i.e. immobilized aminoacylase, for the continuous production of L-amino acids from acetyl-DL-amino acids. This is the first industrial application of immobilized enzymes in the world. Since then we [3, 4, 5, 6, 7] also carried out the industrial applications of immobilized microbial cells for the continuous productions of L-aspartic acid and L-malic acid using immobilized microbial cells with polyacrylamide gel.

For further improvement of these immobilization systems, we investigated many synthetic and natural polymers as a matrix for entrapping enzymes and microbial cells into gel lattice. As a result, we [8, 9] found that "κ-carrageenan" is one of the most suitable polymer for immobilization of microbial cells.

In this paper, the immobilization of microbial cells with polyacrylamide gel and κ-carrageenan, and their industrial application are reviewed.

I. Immobilization with Polyacrylamide Gel and Its Application

1. Production of L-aspartic acid using immobilized *Escherichia coli* [3, 4, 5]

L-Aspartic acid is used for medicines and food additives, and it has been industrially produced by fermentative or enzymic methods from ammonium fumarate using the action of aspartase as shown in the following reaction.

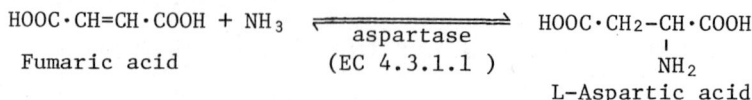

$$HOOC \cdot CH=CH \cdot COOH + NH_3 \underset{\text{aspartase}}{\overset{\longrightarrow}{\rightleftharpoons}} HOOC \cdot CH_2-CH \cdot COOH$$

Fumaric acid   (EC 4.3.1.1) $\overset{|}{NH_2}$

L-Aspartic acid

This reaction has been carried out by batch procedure, which has disadvantages for industrial purpose. Thus, we studied the continuous production of L-aspartic acid using a column packed with immobilized aspartase. As the aspartase is an intracellular enzyme, it was necessary to extract the enzyme from microbial cells before immobilization. Extracted intracellular enzyme is generally unstable, and most of the immobilization methods we tried resulted in low activity and poor yield. Although entrapment into polyacrylamide gel lattice gave relatively active immobilized aspartase, its operational stability was not sufficient, i.e. the half-life was 30 days at 37°C [10]. Therefore, this immobilized aspartase was not satisfactory for the industrial production of L-aspartic acid.

So we considered that if the whole microbial cells could be immobilized directly, these disadvantages might be overcome, and we studied the immobilization of whole microbial cells. Reports on immobilization of whole microbial cells had been very scarce at that time, so we tried various methods for immobilization of E. coli cells. Among the methods tested, the most active immobilized E. coli cells were obtained by entrapping the cells into polyacrylamide gel lattice [3].

An interesting phenomenon was observed with these immobilized cells. When the immobilized E. coli cells were suspended at 37°C for 24-48 hours in substrate solution, its activity increased 10 times higher. The increase of enzyme activity was observed even in the presence of chloramphenicol, inhibitor of protein synthesis. Therefore, this activation was considered not to be the result of protein synthesis but to be due to increased permeability caused by autolysis of E. coli cells in the gel lattice. This was also confirmed by the electron micrographs of immobilized E. coli cells, which indicated that lysis of cells had occurred. Even when lysis of the cells did occur, the aspartase could not leak out from the gel lattice, but the substrate, ammonium fumarate, and the product, L-aspartate, passed easily through the gel lattice.

Using a column packed with immobilized E. coli cells, conditions for continuous production of L-aspartic acid from ammonium fumarate were investigated in detail, and an aspartase reactor system was designed. The system is essentially the same as that for the immobilized aminoacylase system [2]. A solution of 1 M ammonium fumarate (containing 1 mM $MgCl_2$, pH 8.5) is passed

through the immobilized E. *coli* cell column at a flow rate of
space velocity=0.6 hr^{-1} at 37°C. The effluent is adjusted to pH
2.8 and then cooled at 15°C. By this simple procedure pure L-
aspartic acid can be obtained without recrystallization in very
high yield. The activity of immobilized cells is quite stable,
and its half-life is 120 days at 37°C.

This new system had been operating industrially since 1973
in Tanabe Seiyaku Co., Ltd. The overall production cost by this
system was reduced to above 60% of the conventional batch process
using intact cells because of the marked reduction in cost for
the preparation of catalysts and of the reduction of labor cost
by automation. Furthermore, the procedure employing immobilized
cells is advantageous from the standpoint of waste treatment.
This is considered to be the first industrial application of entrapped microbial cells in the world.

2. Production of L-malic acid using immobilized *Brevibacterium ammoniagenes* [6, 7]

In succession to the L-aspartic acid production, in 1974 we
succeeded in the third industrial application, i.e. the production
of L-malic acid from fumaric acid by immobilized microbial cells.
L-Malic acid is an essential compound in cellular metabolism, and
is mainly used in pharmaceutical field. L-Malic acid can be produced by fermentative or enzymatic methods from fumaric acid by
the action of fumarase as follows.

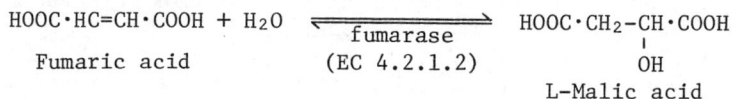

$$HOOC \cdot HC=CH \cdot COOH + H_2O \;\underset{\text{fumarase}}{\overset{}{\rightleftharpoons}}\; HOOC \cdot CH_2-CH \cdot COOH$$

  Fumaric acid   (EC 4.2.1.2)   OH

                                                    L-Malic acid

In this case, reaction reaches an equilibrium when about 80%
of fumaric acid are converted to L-malic acid. We investigated
continuous fumarase reactions using immobilized microbial cells.

Several microorganisms having high fumarase activity were
immobilized by polyacrylamide gel method and their activities
were investigated. B. *ammoniagenes* showed the highest activity
before and after immobilization. However, when immobilized B.
ammoniagenes was used for the production of L-malic acid from
fumaric acid, some by-products were formed. Besides remaining
fumaric acid, considerable accumulation of succinic acid was observed in the reaction mixture. Although fumaric acid can be
easily separated by acidifying the reaction mixture, separation
of succinic acid from L-malic acid is very difficult. Therefore,
the point of success for industrial production of pure L-malic
acid is the prevention of succinic acid formation during the enzyme reaction. So, we tried various methods to suppress succinic
acid formation.

Among the methods tested the detergents used as the solu-
bilizers for membrane- or particle-bound enzymes, bile extract,
bile acid and deoxycholic acid,were found to be effective to de-

crease succinic acid formation in immobilized B. *ammoniagenes*
cells.

Then the conditions for continuous production of L-malic
acid by a column packed with this bile extract treated immobilized
cells was studied. When 1 M sodium fumarate (pH 7.0) is passed
through the column at 37°C at flow rate of space velocity=0.2 hr^{-1},
the reaction reaches an equiliblium. From the effluent of the
column, L-malic acid can be obtained by ordinary methods. Average
yield of pyrogen-free pure L-malic acid from consumed fumaric acid
is around 70% of the theoretical. Tanabe Seiyaku Co., Ltd. is
operating this production system since 1974, and we are satisfied
both with the economical efficiency and with the quality of product.

3. Production of other useful compounds using immobilized cells

Besides these two industrial applications of immobilized
microbial cells above mentioned, we studied efficient continuous
methods for the production of useful organic compounds using immobilized cells as shown in Table 1. Immobilization of respective
microorganisms was carried out by polyacrylamide gel method.

Citrulline [11] used for medicine could be produced from
arginine in higher yield by arginine deiminase activity of *Pseudomonas putida* immobilized with polyacrylamide gel. Half-life of
the column is around 140 days at 37°C.

Also, urocanic acid [12], a sun screening agent in pharmaceutical and cosmetic field, 6-aminopenicillanic acid [13], an important intermediate for synthetic penicillin, NADP [14, 15],
glucose-6-phosphate [16] and glutathione [17, 18] were found to be
prepared by this immobilized cells using polyacrylamide method.

These processes are considered to be more advantageous for
the mass production of respective useful compounds than the batch
method using extracted enzyme or microbial broth.

II. Immobilization with Carrageenan and Its Application

1. New matrix, κ-carrageenan, for immobilization of microbial
    cells

As described above, the polyacrylamide gel method is advantageous for immobilization of microbial cells and for industrial
application. However, there are some limitations in this method.
That is, some enzymes are inactivated during immobilization procedure by the action of acrylamide monomer, β-dimethylamino-
propionitril, potassium persulfate or heat of the polymerization
reaction. Therefore, this method has limitation in application
for immobilization of enzymes and microbial cells. Thus, to find
out more general immobilization technique and to improve the productivities of immobilized microbial cell systems we studied new
immobilization techniques. As the results, we have found out κ-
carrageenan is very useful for immobilization of cells [8]. κ-
Carrageenan, which is composed of unit structure of β-D-galactose

TABLE 1

CONTINUOUS PRODUCTION USING IMMOBILIZED MICROBIAL
CELLS PREPARED BY POLYACRYLAMIDE GEL METHOD

Enzyme system	Microbial cell	Substrate	Product	Operational stability (Half-life, day, Temp.°C)	Reference
L-Arginine deiminase	*Pseudomonas putida*	L-Arginine	L-Citrulline, NH_3	140 (37°C)	11
L-Histidine ammonia-lyase	*Achromobacter liquidum*	L-Histidine	Urocanic acid, NH_3	180 (37°C)	12
Penicillin amidase	*Escherichia coli*	Penicillin	6-APA*	42 (30°C)	13
NAD-Kinase	*Achromobacter aceris*	NAD, ATP	NADP, ADP	20 (37°C)	14
Polyphosphate-NAD kinase	*Brevibacterium ammoniagenes*	NAD, (Pi)n	$NADP, (Pi)n-1$	15 (36.5°C)	15
Polyphosphate-glucose kinase	*Achromobacter butyri*	Glucose, (Pi)n	Glucose-6-phosphate	20 (37°C)	16
Glutathione synthetase	*Saccharomyces cerevisiae*	L-Glutamic acid, L-Cysteine, Glycine	Glutathione	20 (30°C)	17

* 6-Aminopenicillanic acid

sulfate and 3,6-anhydro-α-D-galactose, is a readily available
polysaccharides isolated from seaweeds, and is non-toxic compounds
widely used as food additives.

κ-Carrageenan has characteristics that it becomes gel under
the mild conditions as follows. It becomes gel by cooling as in
the case of agar. Gelation occurs by contact with an aqueous
solution containing metal ions such as alkali metal ions, alkaline
earth metal ions and other bi-or trivalent metal ions. It becomes
gel by contact with an aqueous solution containing ammonium ion or
amines such as aliphatic or aromatic diamines and amino acid derivatives. Gelation also occurs by the contact with water-
miscible organic solvents.

Therefore, taking into consideration of characteristics of
the enzyme-protein and the kind of substrate and product, we can
choose the most suitable procedure for immobilization of the
microbial cells. In our experiences, the procedures cooling
and/or contacting with an aqueous solution containing K^+ or NH_4^+
are very easily carried out for gelation, and as these conditions
are very mild, the immobilized preparation having high activity
can be obtained [9].

The immobilization procedure of microbial cells using κ-
carrageenan is as follows. A cell suspension is warmed at 37°∿
50°C, and κ-carrageenan dissolved in physiological saline is
warmed at 37°∿60°C. The both are mixed, and the mixture is cooled
and/or contacted with an aqueous solution containing a gel inducing agent. After this treatment, the gel is granulated in
suitable particle size, and the immobilized cells can be obtained.
If the operational stability of immobilized cells is not satisfactory, the immobilized cells are treated with hardening agents
such as hexamethylenediamine and glutaraldehyde. As the results,
the stable immobilized cells can be obtained [9].

By this simple procedure many kinds of microbial cells are
successfully entrapped into gel-lattice. Further, anogher advantage of this method using κ-carrageenan is that various shapes of
immobilized preparations, such as cubic, bead and membrane types,
can be easily tailored for particular application purposes. For
instance, bead type can be readily prepared by dropping the mixture of cell suspension and κ-carrageenen solution into stirred
solution containing a gel inducing reagent.

2. Production of L-aspartic acid using immobilized *Escherichia
coli*

Since κ-carrageenan was found to be a useful matrix for immobilization of microbial cells, we extensively studied this
carrageenan method to improve the industrial production of L-
aspartic acid using immobilized *E. coli* cells prepared by polyacrylamide gel method [18].

The aspartase activity and the operational stability of immobilized *E. coli* cells with κ-carrageenan were compared with
that of immobilized one with polyacrylamide. As shown in Table 2,

TABLE 2
HARDENING TREATMENT FOR STABILIZATION OF ASPARTASE
ACTIVITY OF IMMOBILIZED *Escherichia coli*

Hardening treatment			Aspartase activity	Half-life
Reagent and final conc.	Temp.	Time	after activation (unit/g cells)	at 37°C (day)
None			56,340	70
Persimmon tannin (0.02%)	37°	2 hr	54,500	86
GA (mM)	4°	15 min		
2.5			39,690	113
5.0			37,460	240
10.0			28,040	252
HMDA ——— GA	4°	30 min		
(mM) (mM)				
85.0 1.7			50,210	75
85.0 17.0			49,900	108
85.0 85.0			49,400	680
85.0 17.0			36,120	443
Polyacrylamide			18,850	120

GA: glutaraldehyde;   HMDA: hexamethylenediamine

enzyme activity of immobilized preparation with κ-carrageenan was
much higher, but the stability of immobilized preparation with κ-
carrageenan was rather lower than that with polyacrylamide. Thus,
in order to increase operational stability of this immobilized
preparation, hardening treatments were carried out. As shown in
Table 2, aspartase activities of immobilized E. coli cells were
somewhat lowered by this hardening treatment, but their operational stabilities were markedly enhanced. Especially, in the case
of 85 mM hexamethylenediamine and 85 mM glutaraldehyde, the half-
life was extended to 680 days, around two years.

   Thus, we compared the productivity of E. coli immobilized
with polyacrylamide and κ-carrageenan for industrial production
of L-aspartic acid (Table 3). The productivity was calculated
from the equation shown in Table 3. When the productivity of immobilized preparation with polyacrylamide was taken as 100, that
of immobilized cells with κ-carrageenan and hardened with glutaraldehyde and hexamethylenediamine was 1500.

TABLE 3
COMPARISON OF PRODUCTIVITY OF Escherichia coli
IMMOBILIZED WITH POLYACRYLAMIDE AND CARRAGEENAN
FOR PRODUCTION OF L-ASPARTIC ACID

Immobilization method	Aspartase activity (unit/g cells)	Stability at 37°C (half-life, day)	Relative productivity
Polyacrylamide	18,850	120	100
Carrageenan	56,340	70	174
Carrageenan (GA)	37,460	240	397
Carrageenan (GA+HMDA)	49,400	680	1,498

GA: glutaraldehyde; HMDA: hexamethylenediamine

$$\text{Productivity} = \int_0^t E_0 \exp(-kd \cdot t) dt$$

E_0=initial activity, kd=decay constant, t=operational period

   As this κ-carrageenan method is apparently more advantageous
than the polyacrylamide method, we have changed conventional polyacrylamide method to this new κ-carrageenan method for industrial
production of L-aspartic acid. This new method gives us very
satisfactory results.

3. Production of L-malic acid using immobilized Brevibacterium
    flavum
   Succeeding to the production of L-aspartic acid, we tried to
apply this κ-carrageenan method to improve the productivity for
L-malic acid [19].

In this case, we screened again the microorganisms having higher fumarase activity. Compared with previously employed *Brevibacterium ammonigaenes*, *Brevibacterium flavum* was found to have higher fumarase activity and showed higher enzyme activity after immobilization with carrageenan. For industrial purpose, the operational stability is also very important. Thus, the stabilities were compared, and the productivities of immobilized preparations for L-malic acid were calculated.

As shown in Table 4, it is evident that κ-carrageenan method is more advantageous than the conventional polyacrylamide method. So, we changed the polyacrylamide method to the κ-carrageenan method in 1977. This new method also gives us satisfactory result for industrial production of L-malic acid.

TABLE 4

COMPARISON OF PRODUCTIVITY OF *Brevibacterium ammoniagenes* AND *Brevibacterium flavum* IMMOBILIZED WITH POLYACRYLAMIDE AND CARRAGEENAN FOR PRODUCTION OF L-MALIC ACID

Immobi-lization method	B. *ammoniagenes*			B. *flavum*		
	Activity (unit/ g cells)*	Half-life at 37°C (day)	Relative produc-tivity	Activity (unit/ g cells)*	Half-life at 37°C (day)	Relative produc-tivity
Polyacryl-amide	5,800 (60%)	53	100	6,680 (34%)	72	152
Carra-geenan	5,800 (60%)	75	142	9,920 (51%)	70	226

* activity after treatment with bile extract

$$Productivity = \int_0^t E_0 \exp(kd \cdot t) dt$$

E_0=initial activity, kd=decay constant, t=operational period

4. Summary of κ-carrageenan method

Besides these immobilized microbial cells, we applied this carrageenan method to immobilize some kinds of extracted enzymes. These results are summarized and compared with those of immobilized preparations using polyacrylamide (Table 5).

On the point of enzyme activity, all tested microbial cells and enzymes could be immobilized in higher activity by carrageenan method compared with polyacrylamide gel method. Half-life of immobilized preparations with κ-carrageenan was increased by hardening treatment, and it was longer than that of polyacrylamide gel method.

The reason of higher stability of immobilized preparation with κ-carrageenan is not clear, but by a several experiments we found that "κ-carrageenan in liquid-state" does not show any stabilizing effect and "κ-carrageenan in gel-state" is essential for stabilization of enzyme activity. This result suggests that gel-matrix of κ-carrageenan may play an important role for this

TABLE 5
SUMMARY OF ENZYME ACTIVITY AND STABILITY
OF IMMOBILIZED MICROBIAL CELLS AND IMMOBILIZED ENZYMES

Enzyme and microorganism (enzyme)	Enzyme activity (unit/g cells or mg protein)		Operational stability			
			Temp. (°C)	Half-life (day)		
	Polyacryl-amide	Carra-geenan		None	Polyacryl-amide	Carra-geenan
E. coli (Aspartase)	18,900 (73%)	49,400* (49%)	37	10	120	680*
St. phaeochromogenes (Glucose isomerase)	4,160 (37%)	4,310* (59%)	60	–	150	289*
B. ammoniagenes (Fumarase)	5,800 (60%)	5,800 (60%)	37	6	53	75
B. flavum (Fumarase)	6,680 (34%)	9,920 (51%)	37	–	72	70
Aminoacylase	10 (50%)	10 (50%)*	37	–	30	60*
Aspartase	190 (29%)	300 (46%)	37	–	20	–
Glucose isomerase	660 (12%)	4,640* (69%)	60	–	5	120*

* hardening with glutaraldehyde and hexamethylenediamine

stabilization.

Here, we summarize characteristics and advantages of this new immobilization method using κ-carrageenan.

(1) This immobilization method is applicable to many kinds of enzymes and microbial cells.

(2) Activity yield of immobilized enzymes and microbial cells is high.

(3) Various shapes of immobilized enzymes and microbial cells suitable for their application purposes can be easily tailored.

(4) Operational stability of immobilized enzymes and microbial cells is high, especially of that treated with appropriate hardening agents.

(5) Number of living immobilized cells can be easily counted, as entrapped cells are readily converted to cell suspension by removing gel inducing agents.

Thus this κ-carrageenan method is considered to be more advantageous for industrial purpose than the conventional polyacrylamide gel method.

III. Living Immobilized Microbial Cells

Above described continuous enzyme reactions using immobilized microbial cells for production purposes are primarily catalized by a single enzyme and the immobilized cells are in dead state, though the enzyme is in active state. However, many useful compounds especially produced by fermentation method are usually formed by multi-step reactions catalyzed with many kinds of enzymes in living microbial cells. Also, these reactions often require generation of ATP and other coenzymes. If immobilized cells are kept in living state, they may be applied for these multi-enzyme reactions. We applied κ-carrageenan method to the living immobilized yeast cells and bacterial cells for productions of ethanol and L-isoleucine, respectively.

1. Production of ethanol using immobilized yeast cells

The immobilization of yeast, *Saccharomyces carlsbergensis*, with κ-carrageenan was carried out as follows. Without harvesting yeast cells, precultured broth was directly mixed with κ-carrageenan solution and made to immobilized cells of bead type. These gels containing a small amount of cells (3.5×10^6 cells/ml of gel) were incubated in complete medium. After 60 hour incubation, the numbers of living cells in gel increased to 1000 folds that of gel before incubation (5.4×10^9 cells/ml of gel).

Using the column packed with this living immobilized yeast cells, we studied continuous ethanol production. As shown in Figure 1 the steady state of number of living cells and ethanol production was maintained for longer than 30 days. In this operation, some leaking or washing out of cells from gel occured. However, the number of cells in solution was low as $10^{6 \sim 7}$ level. While the cells in gel were maintained at the high levels of 10^9. Therefore,

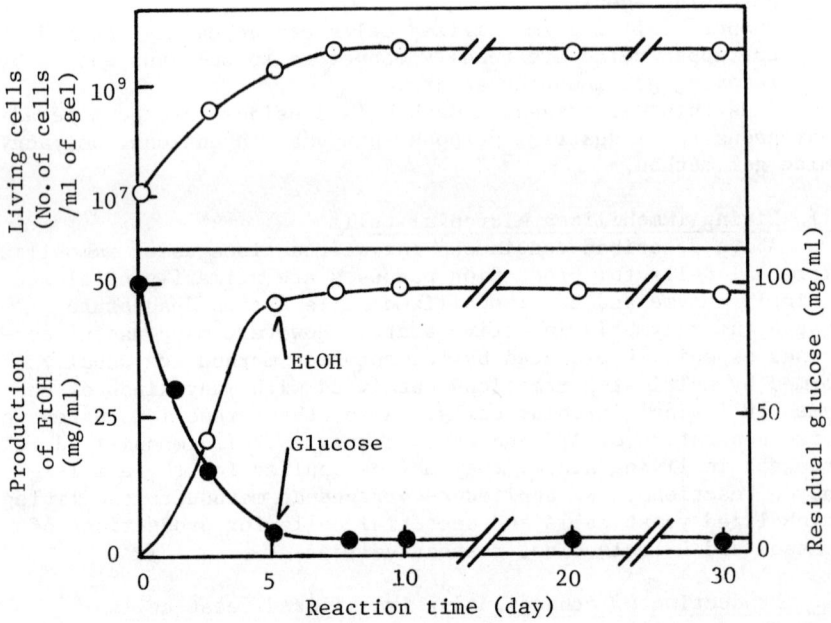

Figure 1. *Pattern of continuous production of ethanol using column packed with immobilized yeast cells. Retention time, 1 hr; gel, 20 mL; flow rate, 20 mL/hr; temperature, 30°C.*

it is clear that almost all alcohol is produced by the cells in
gel. Gel bead also kept its shape after one month operation.
Produced ethanol concentration was 50 mg/ml and residual glucose
concentration was about 2 mg/ml. These values indicate that the
conversion of glucose to ethanol in this system is almost 100% of
theoretical yield even at the fast flow rates such as retention
time is 1 hour. This technique is further being improved, and
now we can produce continuously over 100 mg/ml ethanol. The
details will be published elsewhere in near future. We think our
system using living immobilized yeast cells is very efficient and
promising for production of ethanol.

2. Production of L-isoleucine using immobilized *Serratia marces-*
    cens
    The above ethanol production is based on glycolysis, anaero-
bic multi—step reactions. On the other hand, the living immobilized cell system may also be applied to aerobic multi-step
reactions such as production of amino acids. We are now industrially producing L-isoleucine by conventional fermentation process.
For the purpose to improve this, we have been studying continuous
L-isoleuicne production using the living immobilized microbial
cell system prepared with κ-carrageenan.
    Continuous operations were carried out using packed bed reac-
tor and fluidized bed reactor. Fresh medium was continuously
supplied and reaction mixture was taken out.
    The results showed that immobilized cell system needs to be
operated with fluidized bed under oxygen supply for continuous
L-isoleucine production. While, in the packed bed reactor without
aeration, formation of L-isoleucine rapidly decreased with gradual
decrease of the numbers of living cells.
    Figure 2 shows the typical pattern of continuous L-isoleucine
production with 2-bed reactors. The steady state was maintained
for long periods in 1st and 2nd beds, and L-isoleucine concentration reached to 3.6 mg/ml.
    From these experiments, if improved further, the living im-
mobilized cell system may become promising method not only for
amino acid production, but also for the productions of antibiotics, steroids, and other useful compounds conventionaly prepared
by aerobic and anaerobic fermentations.
    As a conclusion, following advantages of immobilized micro-
bial cells are expected.
  (1) Process for extraction and/or purification of enzyme are not
      necessary.
  (2) Yield of enzyme activity on immobilization is high.
  (3) Operational stability is generally high.
  (4) Cost of enzyme is low.
  (5) Application for multistep enzyme reaction may be possible.
    In enzyme reactions by immobilized microbial cells there are
some limitations as follows.
  (1) When enzymes are intracellular.

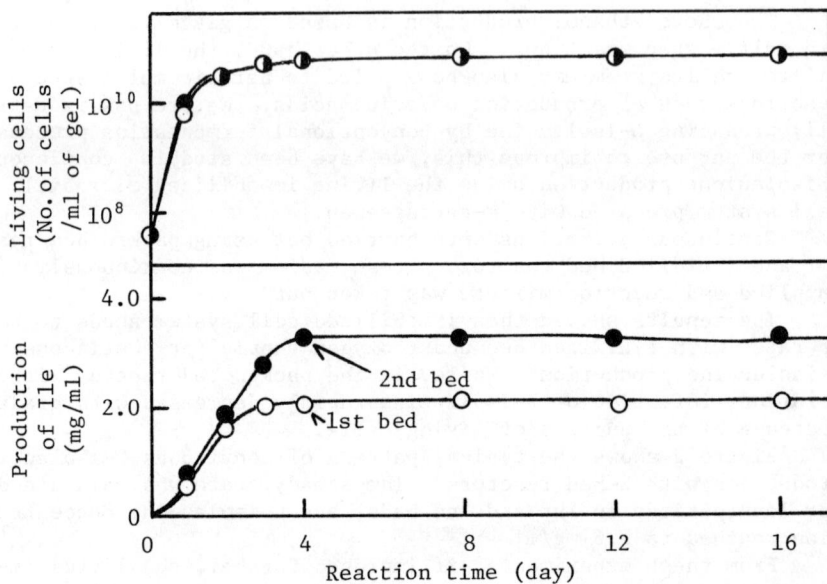

Figure 2. Pattern of continuous production of L-isoleucine using immobilized S. marcescens *cells in two-bed reactor. Retention time, 10 hr + 10 hr; working volume, 150 mL; gel, 50 mL; flow rate, 15 mL/hr; temperature, 30°C.*

(2) When enzymes extracted from cells are unstable during and after immobilization.

(3) When microorganisms contains no interfering enzymes, or interfering enzymes are readily inactivated or removed.

(4) When substrates and products are not high molecular compounds.

In this paper, the application of immobilized microbial cells for chemical process in our Research Laboratory was reviewed. But the technique has been new extending to variety of fields such as analytical, medical, food processing and so on. This is a new and typical multidiscipinary science. For further development of this immobilized systems, I heartily hope, as a biochemist and enzymologist, to cooperate and collaborate with many scientists of varied disciplines, catalytic chemistry, organic chemistry, polymer chemistry and chemical engineering.

Literature Cited

1. Tosa, T., Mori, T., Fuse, N., and Chibata, I., Enzymologia (1966), 31, 214–224.
2. Chibata, I., Tosa, T., Sato, T., Mori, T., and Matuo, Y., "Fermentation Technology Today" (G. Terµi ed.), p.383, Soc. Ferment. Technol., Osaka, Japan (1972).
3. Chibata, I., Tosa, T., and Sato, T., Appl. Microbiol. (1974), 27, 878–885.
4. Tosa, T., Sato, T., Mori, T., and Chibata, I., Appl. Microbiol. (1974), 27, 886–889.
5. Sato, T., Mori, T., Tosa, T., Chibata, I., Furui, M., Yamashita, K., and Sumi, A., Biotechnol. Bioeng. (1975), 17, 1797–1804.
6. Yamamoto, K., Tosa, T., Yamashita, K., and Chibata, I., Europ. J. Appl. Microbiol. (1976), 3, 169–183.
7. Yamamoto, K., Tosa, T., Yamashita, K., and Chibata, I., Biotechnol. Bioeng. (1977), 19, 1101–1114.
8. Takata, I., Tosa, T., and Chibata, I., J. Solid-Phase Biochem. (1976), 2, 225–236.
9. Tosa, T., Sato, T., Mori, T., Yamamoto, K., Takata, I., Nishida, Y., and Chibata, I., Biotechnol. Bioeng., under contribution.
10. Tosa, T., Sato, T., Mori, T., Matuo, Y., and Chibata, I., Biotechnol. Bioeng. (1973), 15, 69–84.
11. Yamamoto, K., Sato, T., Tosa, T., and Chibata, I., Biotechnol. Bioeng. (1974), 16, 1589–1599.
12. Yamamoto, K., Sato, T., Tosa, T., and Chibata, I., Biotechnol. Bioeng. (1974), 16, 1601–1610.
13. Sato, T., Tosa, T., and Chibata, I., Europ. J. Appl. Microbiol. (1976), 2, 153–160.
14. Uchida, T., Watanabe, T., Kato, J., and Chibata, I., Biotechnol. Bioeng. (1978), 20, 255–266.
15. Murata, K., Kato, J., and Chibata, I., Biotechnol. Bioeng., in press.
16. Murata, K., Uchida, T., Kato, J., and Chibata, I., Europ. J.

Appl. Microbiol. Bioeng., in press.
17. Murata, K., Kato, J., and Chibata, I., Europ. J. Appl.
     Microbiol. Bioeng., in press.
18. Nishida, Y., Sato, T., Tosa, T., and Chibata, I., Enzyme and
     Microbiol. Technology, under contribution.
19. Takata, I., Yamamoto, K., Tosa, T., and Chibata, I., Europ.
     J. Appl. Microbiol. Biotechnol., under contribution.

RECEIVED February 15, 1979.

Application of Immobilized Whole Cells in Analysis

BO MATTIASSON

Biochemistry 2, Chemical Center, University of Lund, S-220 07 Lund, Sweden

In the early development of enzyme technology interest was focused mainly on the production and elucidation of pure, well-defined systems, preferably with the enzymes covalently bound to their support (1). Adsorption via charge-charge or via hydrophobic interactions (2,3) soon proved an attractive alternative, especially when crude enzyme preparations were used on a large scale. For them the cost of the support often exceeded that of the enzyme. Another approach adopted was the use of biospecific reversible immobilization (4).

As more and more sophisticated systems were studied the demands placed on the immobilized enzyme preparation constantly increased. The low stability of intracellular enzymes as well as the problems encountered in coenzyme regeneration accelerated the development of immobilized microorganisms or organelles (5). Today more and more of complex biosynthetic reactions are carried out with immobilized whole cells than with immobilized purified enzymes (6,7,8,9). However, in the field of enzymic analysis, where the use of immobilized enzymes is best established, purified enzymes are still used (10,11,12).

Advances expected in immobilization techniques for whole cells and the increasing range of applications of modern enzyme based analyses will lead to a wider use of immobilized organelles and whole cells.

This paper gives a short review of the literature in this field and suggests new possible ways of development of microorganism-based analysis.

Since most of the work done in this area has been carried out on immobilized whole cells, the discussion will be concentrated on this alternative, but naturally, in many cases organelles might be equally good.

Analytical applications involving whole cells or organelles can be based on different prerequisites dictating the choice of the system used.

The immobilized cells may either be applied as alternatives to analysis based on purified enzymes or may fill demands not met by conventional systems.

ANALYSIS INVOLVING LARGE PARTS OF THE TOTAL CELLULAR METABOLISM

The fact that the living cell possess enzyme activities for a broad spectrum of substrates is advantageous when cells are used in qualitative analysis of complex media e.g. measurement of the total content of biodegradable matter in waste waters or when used as poison detectors.

Recording of such undefined reactions, requires the use of transducers registering variables reflecting the total metabolism of the immobilized cell. Such devices are oxygen electrodes (13) for the measurement of total oxygen consumption, biofuel cells for monitoring metabolic events in anaerobic bacteria (14) and calorimeters for registering total heat generated from the integrated metabolism (15,16).

BOD-sensors

The total content of organic material in a waste water is normally determined as BOD (Biological Oxygen Demand). This parameter reflects the amount of oxygen needed for oxidation of all the organic substances present. An assay following conventional procedures takes in the order of 5 days. Serious efforts have been made to shorten this time (17,18,19).

Measurements with oxygen electrodes. The use of oxygen electrodes in combination with immobilized microorganisms has recently been described (20). The bacteria used consisted of a mixed culture isolated from soil and grown on artificial medium before being harvested and immobilized. The preparation was then applied to the system as a gel layer on the bottom of the reaction vessel. The bacterial layer was covered with the water sample to be analyzed, and at the top of the closed vessel an O_2-electrode was immersed in the water. The alternative arrangement using a thin membrane containing microorganisms, wrapped around the sensitive part of the electrode, is shown in fig 1 (21).

The sample solution was saturated with oxygen prior to assay and the electrode was immersed into the reaction mixture. As a result of microbial metabolism, oxygen tension succesively decreased within the membrane until a steady state was reached, i.e. balance between consumption of oxygen and diffusion of oxygen into the gel layer from the surrounding medium. The steady state plateau was read and the amplitude was plotted against the concentration of organic material present in the solution (22). Fig. 2.

The figure shows that the electrode response varies with the increasing amount of organic matter in the solution (the standard solutions used consisted of a mixture of glucose and glutamate). Waste waters with a relatively constant composition of biodegra-

dable organic material were also assayed. Such waters were obtained from a slaughter house, food factories and alcohol factories. The results of these assays were in good agreement with
those obtained with conventional techniques.

Use of biofuel cells. A similar approach was adopted but
with a totally different metabolic procedure. Anaerobic organisms
were used in biofuel cells with electrodes as depicted in fig. 3
(22).
    Clostridium butyricum was immobilized within a thin layer
of polyacrylamide. The electrode was calibrated against glucose--
glutamate in a way similar to that used for calibration of the
oxygen electrode BOD-sensor. The biofuel cell generates a current
by oxidation at the electrode of hydrogen and formate formed in
bacterial metabolism. A good correlation was found between BOD of
the water and the current generated.
    Good agreement was also found between the waste-water BOD as
determined by the present method and conventional procedures.

Thermal analysis using the microbe thermistor. A third
approach to determine the content of organic matter in various
solutions used the microbe thermistor (15). We used a simple
semiadiabatic flow microcalorimeter, when applied in enzyme
analysis, usually called "enzyme thermistor" (Fig. 4.) (12). The
microcolumn was filled with polyacrylamide entrapped yeast cells
(Saccharomyces cerviseae). The system was continuously perfused
with buffer and small amounts of substrate were introduced as
pulses into the flow by means of a three-way valve. Passage of
the flow through the bio-bed, is accompanied by a reaction heat
(almost all enzyme-catalyzed reactions are accompanied by heat
evolution). The heat is transported with the flow and out of the
column. At the outlet, a thermistor measures any change in temperature. Such changes are amplified and recorded. The model
studies performed showed that glucose was easily quantified and
that more complex molecules, such as casein or amylose, after
enzymic pretreatment with pronase and amylase respectively, also
could be measured. Thus, provided a suitable culture of microorganisms is used in the reaction bed, the microbe thermistor
should prove a useful tool for BOD-analyses, for example.

Poison guards

    Immobilized whole cells are very suitable as sensors in
environmental control instruments. In multipolluted waste water
it might be difficult to assess the concentrations of each of the
polluting molecular species. This is of less interest. Of far
greater interest is the net effect on the environment. This point
can be cleared up, at least partly, by testing a small sample of
the polluted water on a biosystem such as a preparation of
immobilized cells.

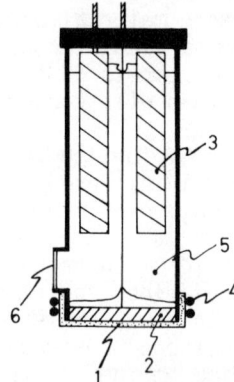

Figure 1. Schematic of the microbial electrode. 1, Clostridium butyricum; 2, platinum electrode; 3, silver peroxide electrode (Ag$_2$O$_2$); 4, O-ring; 5, electrolyte (0.1M phosphate buffer); 6, anion exchange membrane (21).

Journal of Solid-Phase Biochemistry

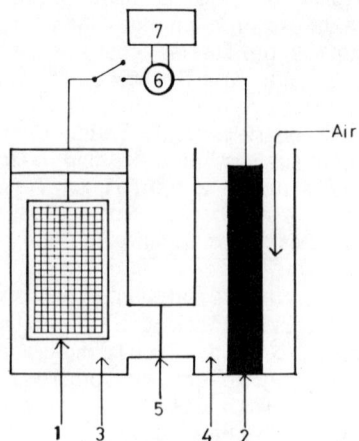

Figure 2. Relationship between the steady state current and the BOD of a standard solution. A solution of 200 mL of 0.1M phosphate buffer containing various concentrations of glucose (2.5–400 mg) and glutamate (2.5–400 mg) was used and current was determined 40 min after insertion of the electrode (21).

Journal of Solid-Phase Biochemistry

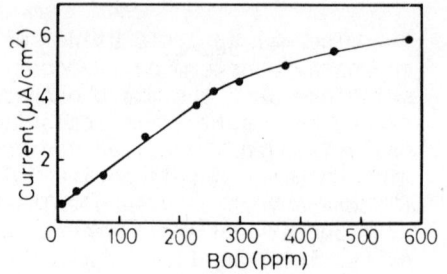

Figure 3. Scheme of BOD sensor. 1, Microbial electrode; 2, carbon electrode; 3, sample waste water; 4, catholyte (0.1M phosphate buffer); 5, anion exchange membrane; 6, ammeter; 7, recorder (14).

Biotechnology and Bioengineering

Immobilized yeast cells were exposed to uncouplers (2,4-dinitrophenol) as well as to arsenate. It is clear from the thermogram in Fig. 5 that these two poisons had opposite effects on the metabolism of the cell (as judged from the heat generated). Arsenate suppressed the generation of heat, while DNP liberated more energy as heat (less metabolic energy conserved as ATP). The inhibiting effect of various poisons can thus be followed with the aid of calorimetry, but only when the toxicity of the substances present is known. can it be used for screening purposes.

The total cell metabolism is more difficult to inhibit than individual activities of purified enzymes. Thus, for rapid, sensitive analysis of discrete samples, analyses based on separate enzymes are preferable, but immobilized cells might prove useful for continuous analysis e.g. as poisons guards, especially in crude solutions of non-predictable composition.

An instrument operating along these lines is a continuous monitoring system for acute toxicity of waste waters (23). This instrument consists of a pump for pumping the waste water, an oxygenator, a bio-step containing a submerged biological filter with rotational plates covered with a thin film of living microorganisms and in the effluent from this unit an O_2-sensor. Using this unit it was possible to register e.g. suddenly occuring Cu^{2+}-concentrations, CN^--concentrations or sudden pH-changes. Fig. 6 shows the response of the system when exposed to various concentrations of CN^-. The poison guards operate in such a way that when a certain level of O_2 (higher than the normal level) is reached in the effluent, i.e. a decrease in metabolic activity is found in the bio-reactor, an alarm is started. After the poison has been eliminated from the system, the bio-bed needs time to recover (e.g. in the example given in Fig. 6 it requires 38 minutes to recover from 5 mg CN/l and 3 hours to recover from 20 mg/ml).

Analysis of antibiotics

In recent years several reports have been published (24,25, 26) on microcalorimetric studies of the action of antibiotics on various microorganisms. The assay principle is that the total heat generated by the cells when fed an appropriate medium is compared with that registered when varying concentrations of the actual antibiotic are supplied.

The conventional calorimetric procedure involves the use of free living cells (24,25,26). As the inhibitory effect of the antibiotic reduces the metabolic activity of the cells it is advantageous to use rather high concentrations of cells. Since one of the characteristics of immobilized systems is that very high concentrations of the catalytically active substances can be used one would expect systems based on immobilized cells to be useful for determining concentrations of antibiotics. Preliminary

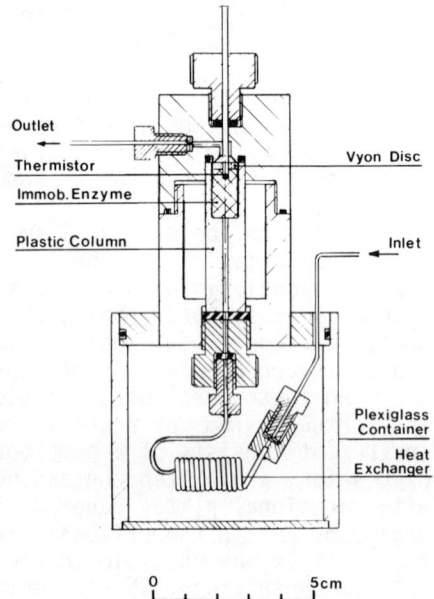

Figure 4. A generalized picture of the
enzyme thermistor (12)

Pure and Applied Chemistry

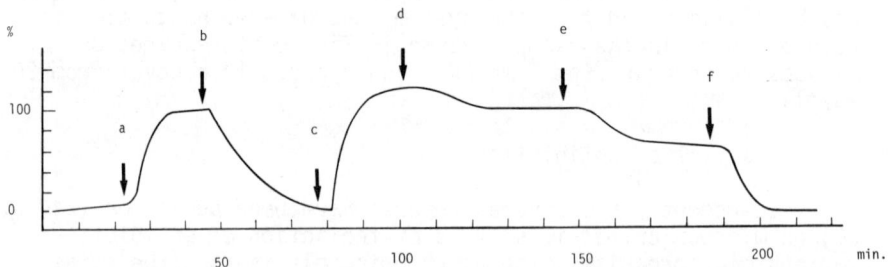

Nature

Figure 5. Thermogram obtained with the microbe thermistor following introduction of various metabolites and inhibitors dissolved in 0.1M potassium phosphate buffer pH 7.0 into the flow. The arrows indicate addition of a, 1mM glucose; b, buffer; c, 1mM glucose + 1mM 2,4-dinitrophenol; d, 1mM glucose;
e, 1mM glucose + 2mM arsenate; f, buffer. The steady-state response to 1mM
glucose is set as 100% (15).

experiments with the microbe thermistor have also confirmed that
they are (27). Thus the antibiotic Nystatin has been determined
in the concentration range 5 - 10 mg/l utilizing immobilized
Saccharomyces cerviseae as sensor.

An alternative method for assay of antibiotics is to use
immobilized cells (28) or purified enzyme (12) for specifically
converting the antibiotic. Both these two methods have been
applied in the assay of cephalosporins.

Assay of substances specifically stimulating the total cell metabolism

When performing bioassays mutants lacking the ability to
produce a certain compound are generally used for assay of the
compound. However, strains of organisms with normal genetic
properties might be used in some cases. Such an example is the use
of Saccharomyces cerviseae for the bioassay of thiamine (29). The
yeast cells were kept on a thiamine limiting medium for a definite
period before being exposed to full medium with known concentrations of thiamine. The stimulation of the metabolic activity was
reflected in cell growth and division which were assayed.

Immobilized Saccharomyces cerviseae cells were placed close
to a O_2-sensitive electrode and the background metabolic activity
was measured in limiting medium with various known concentrations
of thiamine. The registered oxygen consumption was correlated with
the thiamine concentration within a concentration range of 0.05 -
2.5 μM (30).

Bioassay of specific substances with the use of immobilized mutants.

The extreme sensitivity reported in conventional bioassays
is due to the fact that in such assays the whole metabolism of the
cell is dependent upon the presence of the substance to be assayed.
Cell cultures grow slowly - a bioassay takes some 48 hours before
the size or the number of colonies can be measured.

An initial step in this normal procedure (which eventually
leads to conditions that are possible to measure) must be changes
in metabolic activity. By combining immobilized organisms with
specific detectors it is possible to detect such changes long
before any macroscopic change takes place.

Immobilized Lactobacillus arabinosus strain ATCC 8014 was
used for analysis of nicotinic acid (31). The vitamin assay was
performed with a pH-electrode and bacteria immobilized in agar.
The assay was based on the well-known fact that Lactobacillus
produces large amounts of lactic acid. Since the metabolism of
such a bacterial preparation is dependent upon the concentration
of nicotinic acid in the medium, the assay could be performed by

correlating the concentration of lactic acid produced with the
concentration of nicotinic acid present. A calibration curve is
shown in Fig. 7. An assay using this technique took approximately
one hour.

ASSAY OF SINGLE SUBSTRATE SPECIES.

When immobilized whole cells were first introduced many
researchers hesitated to use them because of their non-speci-
ficity and the expected by-product formation. Efforts have been
made to utilize as many as possible of the potential advantages
offered by immobilized whole cells, and at the same time to make
the biocatalyst specific. In such endeavours selective enzyme
denaturation, cell membrane modification (9,32,33) and specific
mutants for example, have been used.
      The literature on processes based on immobilized whole
cells contains very little about by-product formation and virtually no analytical data on the final product. In large scale
processes, by-products can be eliminated in a subsequent purification step, but this is not possible in the analytical field
since the analytical answer is normally directly correlated with
the metabolic activity of the immobilized preparation with the
result that in the event of side-reactions the figures given by
the analysis will be too high.
      This restricts the use of immobilized whole cells in ana-
lysis of single substances to three main applications: a) determination of a substance in a pure product stream when production
of by-products does not matter as long as that part of the total
metabolism involved is constant, thus giving a reproducible
signal; b) use of organisms with increased amounts of a specific
enzyme or c) mutants that have completely lost one enzyme function.

      Using the microbe thermistor (15) it was possible to deter-
mine glucose concentrations in the perfusion medium. Fig. 8 . The
response was slower than is the case when purified enzyme preparations were used (34).
      Fig. 9 gives the curve for Δt versus glucose concentration.
In this example, pure substrate solutions were used and since
the test was carried out under aerobic conditions an essential
part of the metabolic system in the yeast cells was used.
      Another example deals with amino acid analysis using immo-
bilized specific microorganisms in combination with selective
electrodes (35). Thus, glutamine could be analyzed by an electrode consisting of a potentiometric ammonia gas sensor and a
layer of the bacterium Sarcina flava (American type culture
collection 147) trapped in the volume between a NH_3-permeable
membrane on the surface of the electrode and a dialysis membrane
in contact with the surrounding solution (Fig. 10). Using this
electrode, steady state potentials were reached within 5 minutes.

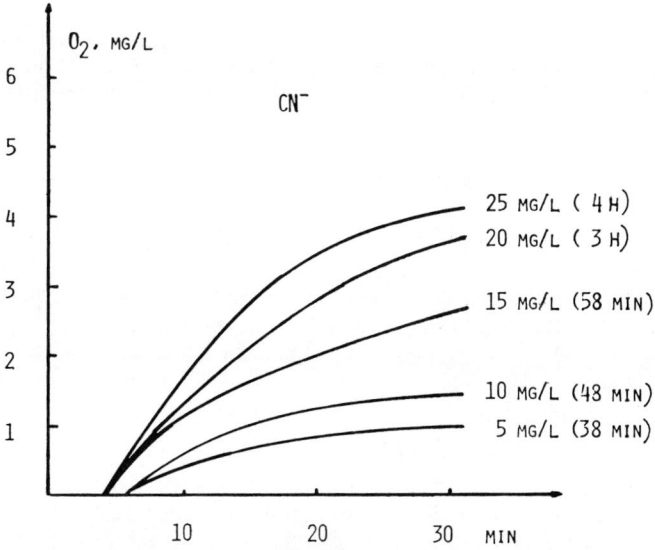

Vatten

Figure 6.   *Toxicity indication for cyanide from the continuous monitoring system upon administration of cyanide to the system (23)*

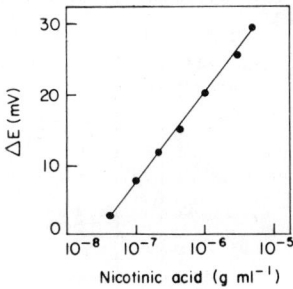

Analytica Chimica Acta

Figure 7.   *Calibration curve for nicotinic acid under optimum conditions (31)*

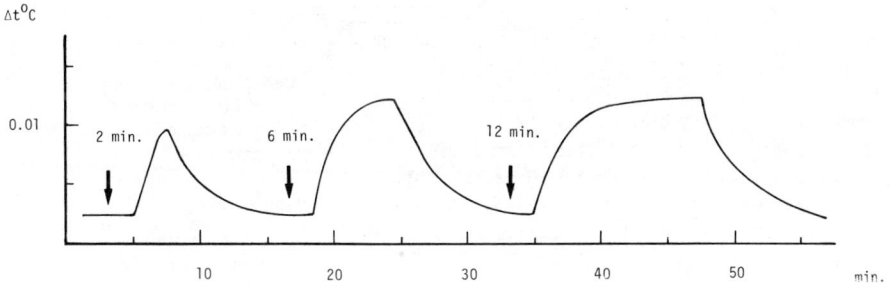

Nature

Figure 8.   *Experimental curves obtained after injection of 1mM glucose pulses (in 0.1M potassium phosphate pH 7.0) of varying durations. Arrows indicate substrate introduction (15).*

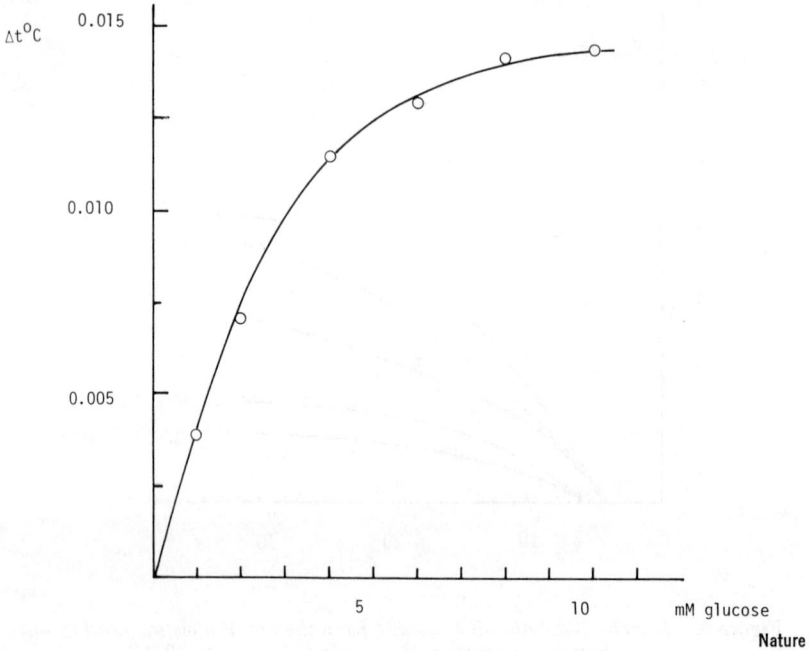

Nature

Figure 9.   Measured peak height ($\Delta t°C$) as a function of glucose concentration. The glucose substrate was dissolved in 0.1M potassium phosphate and injected as 1-min pulses (15).

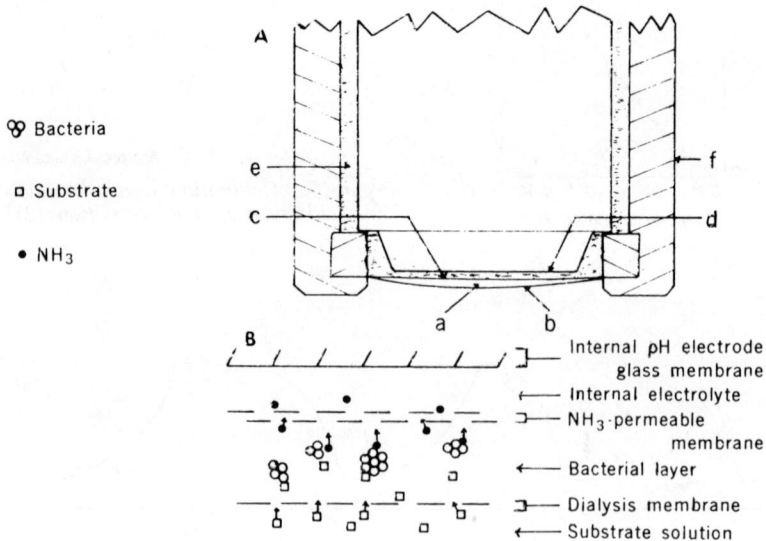

Science

Figure 10.   (A) Schematic diagram of the bacterial electrode: a, bacterial layer; b, dialysis membrane; c, gas-permeable membrane; d, internal sensing element; e, internal filling solution; and f, plastic electrode body. (B) Detail of the membrane phases (35).

The linear range 10^{-4} - 10^{-2} M (Fig. 11) is far better than what
is reported for electrodes based on purified glutaminase (E.C.
3.5.1.2.) (36).
Furthermore, the response of this electrode was not dis-
turbed when moderate concentrations of other amino acids were
present. When determining glutamine in serum, a calibration curve
with the same slope as of that in buffer was obtained, but owing
to differences in viscosity of the sample from that of standard
solutions, there was a slight shift along the potential axis.
Similar electrodes were also reported for the assays of
cystein,aspartate and arginine (37,38).
The stability of the devices described here is very good,
e.g. the glutamine sensor was used for at least two weeks. In
addition, the electrode could afterwards be reactivated by exposing the bacteria electrode unit to nutrients for a short
period. Following such treatment, operational characteristics
almost identical with the native bacteria electrode could be
measured.

Bakers yeast (Saccharomyces cerviseae) has also been used
in combination with mass spectrometry thereby assaying volatile
products from cell metabolism (39). The basic concept here is
that the living cells are kept in a flow system continuously fed
with buffer or substrate and that any volatile reactant passes
through a semipermeable membrane into the evacuated room containing an ionizer and a quadrepole mass spectrometer (Fig. 12)
(40).
Using this technique on immobilized enzymes substrate
concentrations down to 10^{-6} - 10^{-5} M can could be determined. As
the assay principle is applicable in all cases where volatile
products are formed, one might imagine a wide range of applications of the technique (40).
Free yeast cells were introduced into the flow system
together with glucose and the volatile products carbon dioxide
and ethanol were analyzed. In this way a linear calibration curve
(ethanol count vs cell number) over the range 10^7 - 10^9 cells was
obtained.
Using cells entrapped in polyacrylamide, glucose could be
determined. However, severe diffusional restrictions of the
system resulted in rather low sensitivity of the system. In this
method as well as in the one using a microbe thermistor (15) ,
better analytical resolution would be obtained provided the
immobilization techniques used are satisfactory.

IMMOBILIZED BACTERIA USED AS SORBENTS IN ANALYSIS

The cell surface is a subject of increasing attention. Many
specific properties of various cells can be explained by the
presence of specific molecules on the cell surfaces. Bacteria

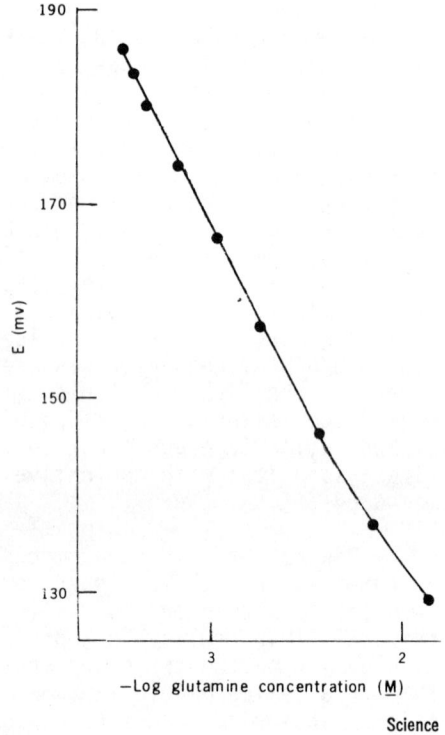

Figure 11. Response of the bacterial sensor to glutamine in reconstituted control serum (1:5 dilution) (35).

Science

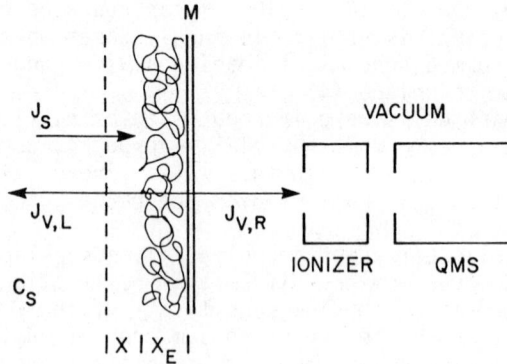

Plenum Press

Figure 12. Basic configuration used for the mode that assays a nonvolatile substrate concentration, C_s, by means of a volatile product. M is the semipermeable membrane, with an enzyme layer of thickness X_E and an assumed aqueous unstirred layer of thickness X on the left. The vacuum of the mass spectrometer (here a quadrupole mass spectrometer (QMS) shown with its ionizer) is to the right of the membrane. J_s is the steady-state substrate flux to the enzyme layer, while $J_{v,r}$ and $J_{v,l}$ are the steady-state right- and left-going volatile product fluxes (40).

possess cell walls which have been studied extensively since a
crucial point in the treatment of bacterial infection has been
to prevent cell wall formation.

Many bacteria possess specific properties because specific
proteins or glycoproteins are situated on the cell surface.
Staphylococcus aureus is an example of bacteria carrying specific
molecules - called protein A (40,41). This protein has specific
binding properties since it binds immunoglobulin subgroups I, II
and IV via their F_c fragments. Protein A, purified as well as
when bound to the cell walls of killed cells, have been used in
radio immunoassay (42,43) and in enzyme immunoassay (44).

Since many bacteria possess specific binding proteins (45)
also other applications of specific binding may be expected.

Immobilized cell walls can also be used for studying and
perhaps also for quantification of viruses. Thus, using columns
with cell walls of Escherichia coli immobilized on kiselguhr it
was possible to study the adsorption of specific phages to the
cell walls as well as the DNA-injection process (46).

IMMOBILIZED ANIMAL CELLS

The immobilization of animal cells is still in its infancy.
This is at least partly because of the fact that such cells are
far more labile and sensitive than microbial cells and that most
immobilization procedures hitherto used have been too drastic for
the animal cells to survive. However, the biospecific interactions offered by glucoproteins on the cell surface and lectins
on the support may be utilized for immobilization to take place
(4,47,48).

Lectin-sorbents can either be used for affinity chromatographic separation of cells (47,48) or for keeping the cells
immobilized in a reversible manner when used for applications
purposes, for example (4).

Red blood cells bound to Concanavalin A - Sepharose or Lens
culinaris lectin - Sepharose was used in combination with immobilized glucose oxidase (E.C. 1.1.3.4) in the thermistor system.
The red blood cells operated as oxygen supply in the reactor bed
leading to better reaction conditions for the oxidase - thereby
widening the practical concentration range for glucose analysis
(4) (Fig. 13).

CONCLUDING REMARKS

Immobilized cells can either, as is demonstrated by the
examples given above, be used as sensors in analytical systems or
by being kept in close proximity to a transducer they can be

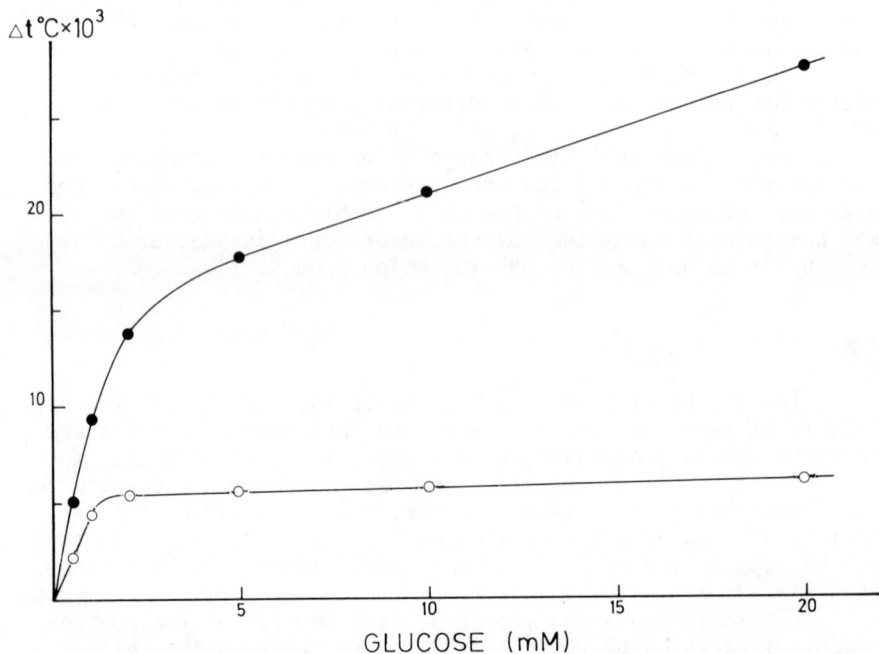

FEBS Letters

Figure 13.   Measured peak height (Δt °C) obtained from a glucose oxidase–con-canavalin A–Sepharose column (○) or a glucose oxidase–red blood cell–concana-valin A–Sepharose column (●) as a function of the concentration of glucose dis-solved in 0.1M Tris–HCl buffer pH 7.0, being 1M in NaCl, 1 mM in MgCl$_2$, MnCl$_2$ and CaCl$_2$ (1 min pulse, flow rate 0.75 mL/min) (4).

studied under well defined conditions. Thus, as is illustrated by the section on poison - guards, immobilized cells may be used to evaluate the effect a certain substance or group of substances may exert on living cells. This does not only include poisons, but also other molecular species such as hormones, antibiotics and drugs etc.

In general, the technique for handling of immobilized cells both easily and specifically and, under well controlled conditions, for changing their milieu offers good possibilities to study cells, cell metabolism, cell physiology and cell toxicology under much more well defined conditions than hitherto.

The first examples of immobilized microorganisms are discussed above and it can be predicted that when the technique for immobilizing mammalian cells developed further, systems analogous to those discussed in this paper may be used for studying basic cell metabolic phenomena as well as in sensitive biospecific analytical systems.

LITERATURE CITED

1. Porath, J. and Axén, R., in Methods in Enzymology, vol. 44,
   (ed. Mosbach, K.) pp 19-45, Academic Press, New York, 1976.

2. Messing, R.A., ibid pp. 148-169.

3. Caldwell, K., Axén, R. and Porath, J., Biotechnol. Bioeng.
   (1976) 18, 433-438.

4. Mattiasson, B. and Borrebaeck, C., FEBS Lett.(1978),85, 119-
   123.

5. Jack, T.R. and Zajic, J.E., in Advances in Biochemical
   Engineering vol. 5, pp 125-145, Spriger Verlag, Berlin, 1977.

6. Ohlson, S., Larsson, P.-O. and Mosbach, K., Nature (1976)
   263, 796-797.

7. Venkatasubramanian, K., Constantinides, A. and Vieth, W.,
   in Enzyme Engineering vol. 3, eds. Pye, E.K. and Weetall,H.H.,
   pp 29-41, Plenum, New York 1978.

8. Klein J., Wagner, F., Washausen, P., Eng, H. and Martin,C.K.A.,
   preprint First Eur. Congress Biotechnol. (Interlaken, Switzer-
   land, Aug. 25-29, 1978) pp 190-193, DECHEMA, Frankfurt/M.,
   W. Germany, 1978.

9. Chibata, I., Tosa, T. and Yamamoto, K., in Enzyme Engineering
   vol. 3, eds. Pye, E.K. and Weetall, H.H., pp 463-468,
   Plenum, New York, 1978.

10. Guilbault, G.G., Handbook of Enzymatic Methods of Analysis,
    pp 445-543, Marcel Decker Inc. New York, 1976.

11. Bowers, L.D. and Carr, P.W., Anal. Chem.(1976) 48, 544-549.

12. Danielsson, B., Mattiasson, B. and Mosbach, K., Pure and
    Appl. Chem.in press.

13. Fatt, I., Polarographic Oxygen Sensors, CRC-Press, Cleveland,
    Ohio,1976.

14. Karube, I., Matsunaga, T., Tsuru, S., and Suzuki, S.,
    Biotechnol. Bioeng.(1977), 19, 1727-1733.

15. Mattiasson, B., Larsson, P.-O. and Mosbach, K., Nature
    (1977), 268, 519-520.

16. Spink, C. and Wadsö, I., Methods Biochem. Anal.(1976),23,1-
    159.

17. Leblance, P.J., J.Water Pollut. Control Fed.(1974),46, 2202.

18. Busch, A.W. and Mirick, N., J. Water Pollut. Control Fed.
    (1962) 34, 354.

19. Mullis, M.K. and Schroeder, E.D., J.Water Pollut. Control
    Fed. 43, 209-215.

20. Karube, I., Mitsuda, S., Matsunaga, T. and Suzuki, S.,
    J. Ferment. Technol.(1977), 55, 243-248.

21. Karube, I., Matsunaga, T. and Suzuki, S., J.Solid-Phase Bio-
    chem.(1977), 2, 97-104.

22. Karube, I., Matsunaga, T., Mitsuda, S. and Suzuki, S.,
    Biotechnol. Bioeng. (1977), 19, 1535-1547.

23. Solyom, P., Vatten (1976) 192-199.

24. Ljungholm, K., Wadsö, I. and Mårdh, P.-A., J.Gen. Microbiol.
    (1976), 96, 283-288.

25. Mårdh, P.-A., Ripa, T., Andersson, K.-E. and Wadsö, I.,
    Antimicrobial Agents and Chemother. (1976), 10, 604-609.

26. Beezer, A.E., Chowdhry, B.Z., Newell, R.D. and Tyrrell,
    H.J.V., Anal.Chem.(1977) 49, 1781-1784.

27. Mattiasson, B., to be published.

28. Suzuki, S. and Karube I., preprint First. Eur. Congr. Bio-
    technol. (Interlaken, Switzerland, Sept 25-29, 1978)
    pp 31-34, DECHEMA, Frankfurt/M, W. Germany. 1978.

29. Witt, I. and Neufang, B., Biochim. Biophys. Acta (1970),
    215, 323-332.

30. Mattiasson, B. and Larsson, P.-O., to be published.

31. Matsunaga, T., Karube, I. and Suzuki, S., Anal. Chim. Acta
    (1978), 99, 233-239.

32. Shimizu, S., Morioka, H., Tani, Y. and Ogata, K., J.Ferment.
    Technol.(1975), 53, 77-83.

33. Ohlson, S., Larsson, P.-O. and Mosbach, K., Biotechnol.
    Bioeng.(1978), 20, 1267-1284.

34. Danielsson, B., Gadd, K., Mattiasson, B. and Mosbach, K.,
    Clin. Chim. Acta (1977), 81, 163-175.

35. Rechnitz, G.A., Riechel, T.L., Kobos, R.K. and Meyerhoff,
    M.E., Science,(1978), 199,440-441.

36. Guilbault, G.G. and Shu, F.R., Anal. Chim. Acta (1971), 56,
    333-338.

37. Rechnitz, G.A., paper presented at the 30th Annual Summer
    Symp. on Analytical Chemistry, Amherst, Mass., U.S.A.
    13 June 1977.

38. Jensén, M.A. and Rechnitz, G.A., personal communication.

39. Weaver, J.C., Reames, F.M., DeAlleaume, L., Perley, C.R.
    and Cooney, C.L., in Enzyme Engineering vol. 4, eds. Broun,
    G.B., Manecke, G. and Wingard, L.B. jr., pp 403-404,
    Plenum, New York, 1978.

40.   Weaver, J.C., in Biomedical Applications of Immobilized
       Enzymes and Proteins, vol. 2, ed. Chang, T.M.S., pp 207-225,
       Plenum, New York, 1977.

41.   Björk, I., Petersson, B.-A. and Sjöqvist, J., Eur. J.
       Biochem.(1972), 29, 579-584.

42.   Jonsson, S. and Kronvall, G., Scand. J. Immunol. (1972), 1,
       414-415.

43.   Brunda, M.J., Minden, P., Sharpton, T.R., McClutchy, J.K.
       and Farr, R.S., J. Immunol.(1977), 119, 193-198.

44.   Mattiasson, B. and Borrebaeck, C., in Proc. Int. Symp.
       Enzyme Labelled Immunoassay of Hormons and Drugs, ed. Pal,S.,
       W. de Gruyter, Berlin, in press.

45.   Myhre, E.B. and Kronvall, G., Infection and Immunity (1977)
       17, 475-482.

46.   Labedan, B., Virology (1978), 85, 487-493.

47.   Rutishauser, U. and Edelman, G.M., in Concanavalin A as a
       Tool, eds. Bittiger, H. and Schnebli, H.P., pp 447-465,
       John Wiley & Son, New York, 1976.

48.   Kinzel, V., Kübler, D., Richards, J. and Stöhr, M.,
       ibid. pp 467-478.

RECEIVED February 15, 1979.

Microbial Electrode Sensors for Cephalosporins and Glucose

SHUICHI SUZUKI and ISAO KARUBE

Research Laboratory of Resources Utilization, Tokyo Institute of Technology, 4259 Nagatsuta-cho, Midori-ku, Yokohama 227, Japan

Determination of organic materials such as whole cells, nutrients and products in fermentation broths is essential for efficient control of fermentation processes.

Electrochemical monitoring of these compounds has definite advantages. For example, wide concentration ranges are measurable without dilution simply by scale switching and the test sample does not need to be optically clear. Many of these methods have utilized enzyme-catalyzed reactions because of the specificity of such reactions. Many reports on applications of enzyme electrodes in clinical and food analysis have been published (1,2). However, enzymes are generally expensive and unstable. Recently many methods have been developed for immobilization of whole cells (3). Immobilized whole cells have been applied to electrochemical sensors. Microbial electrodes using immobilized whole cells and electrochemical devices have been developed by the authors (4,5,6). Such microbial electrodes can be classified under three categories as shown in Figure 1.

Bacterial electrodes using whole cell suspensions and an ammonia sensing electrode also have been reported by Rechnitz et al (7,8).

Microbial electrodes for cephalosporins, glucose and other materials are described in this paper.

Microbial Electrode Sensor for Cephalosporins

The determination of antibiotics in fermentation broths is important for control of an antibiotic fermentation. Antibiotics are usually determined by microbioassay based on turbidimetric or titrimetric methods. However, these methods require a long time for cultivation of bacteria (9). Therefore, rapid and continuous determination of antibiotics produced is difficult by the micro-bioassay.

A microbial electrode consisting of a bacteria-collagen membrane reactor and a combined pH electrode for cephalosporins in a fermentation broth are described here. It was found that

MICROBIAL ELECTRODE

1. <u>Amperometric determination of microbial respiration</u>
   Immobilized whole cells + Oxygen electrode

   a) Glucose
   b) Assimilable sugars
   c) Acetic acid
   d) Alcohol
   e) Biochemical oxygen demand (BOD)

2. <u>Amperometric determination of electroactive</u>
   <u>substances</u>
   Whole cells (or immobilized whole cells) +
   biochemical fuel cell

   a) Cell Population
   b) Vitamin B_1
   c) BOD (immobilized whole cells)

3. <u>Potentiometric determination of electroactive</u>
   <u>substances</u>

   a) Immobilized whole cells + pH electrode
     i) Cephalosporins
    ii) Nicotinic acid
   b) Immobilized whole cells + CO_2 electrode
      glutamic acid

Figure 1.   Classification of microbial electrodes

Citrobacter freundii produced cephalosporinase. Cephalosporinase catalyzed the reaction shown in Figure 2 and hydrogen ions are liberated from the cephalosporinase reaction. Cephalosporins may be determined from proton concentration in a medium by using immobilized cephalosporinase and a pH electrode. However, immobilization of cephalosporinase is difficult because the molecular weight of cephalosporinase is low (MW 30,000) and the enzyme is unstable. Therefore, whole cells of Citrobacter freundii were immobilized in collagen membrane.

Apparatus and Procedures. Citrobacter freundii B-0652 was employed in our experiments. The bacteria were cultured under aerobic conditions at 37°C for 5 h in 1000 ml of the heart infusion broth (pH 7.0). The cells were centrifuged at 5°C and 8000 g and washed three times with deionized water. About 4.2 g of wet cells were obtained from the broth.

Collagen fibril suspension was prepared as described previously (10). Wet cells (4 g) were added to 60 g of 0.75 % collagen fibril suspension. The bacteria-collagen membrane was prepared by casting the suspension on a Teflon plate and drying it at room temperature for 20 h . The bacteria-collagen membranes were treated with 1 % glutaraldehyde solution (pH 7.0) for 1 min and dried again at room temperature. The thickness of the bacteria-collagen membrane was 50-60 μ. The activity yield of cephalosporinase was about 9 %. However, cephalosporinase in whole cells immobilized in collagen membrane was stable.

The system used for continuous determination of cephalosporins is shown in Figure 3. The reactor was a biocatalytic type (11) (acryl plastic, diameter 1.8 cm, length 5.2 cm) with a spacer (glass rod, diameter 1.4 cm, length 5.0 cm) located in the center. The inner volume of the reactor was 4.1 ml. The bacteria-collagen membrane (10 x 5.5 cm^2 53000 I.U.) was rolled with plastic net (5 x 20 cm^2, 20 mesh) and inserted into the reactor.

Phosphate buffer (0.5 mM, pH 7.2) was transferred continuously to the reactor and sensing chamber. Then, 10 ml of sample solution containing various amounts of cephalosporins was transferred to the reactor at a flow rate of 2 ml/min by a peristaltic pump and the hydrogen ion concentration in the sample solution was determined by a combined pH electrode (GC-125 C, TOA Electronics Co. Tokyo) and displayed on a recorder (Model CDR-11A, TOA Electronics Co.).

Cephalosporinase activity of the membrane was also determined independently by the method of Perret (12).

Response Curve of the Electrode. From Nernst's formula, the membrane potential of the glass electrode is proportional to the logarithm of the proton concentration. Protons are liberated from the cephalosporinase reaction. Therefore, hydrogen ion concentration in a sample solution measured by the pH electrode is proportional to the logarithm of the cephalosporin concentration

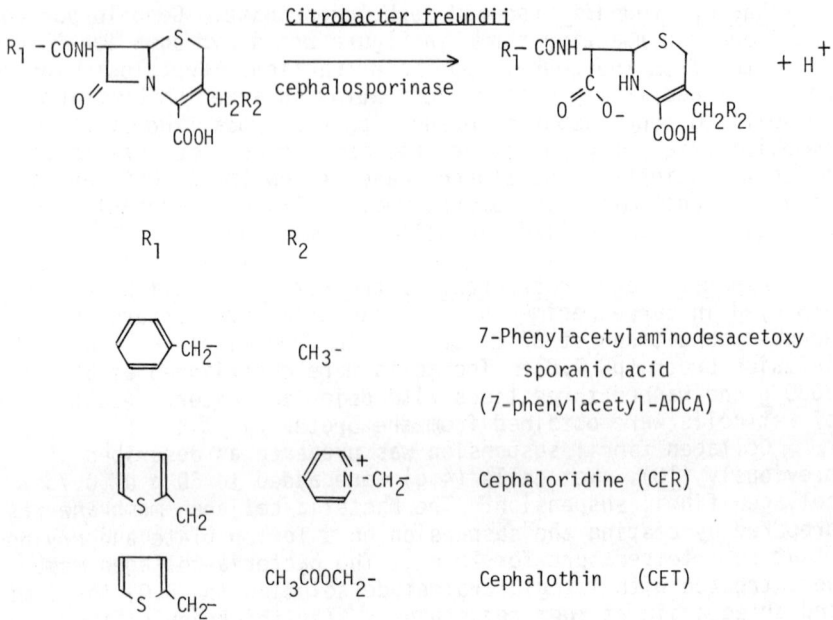

Figure 2.   Cephalosporinase reaction

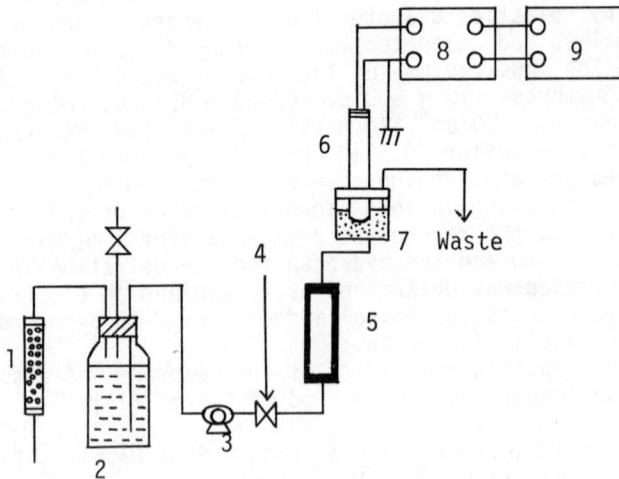

Figure 3.   Scheme of the microbial electrode sensor for cephalosporins. 1. Soda lime. 2. Buffer reservoir. 3. Peristaltic pump. 4. Inlet of the sample. 5. Immobilized whole cell reactor. 6. Combined glass electrode. 7. Sensing chamber. 8. Amplifier. 9. Recorder.

Sample solutions (10 ml) containing various amounts of
phenylacetyl-7ADCA were transferred to the reactor and the
potential in the sensing chamber gradually increased with time
until a maximum was reached. The response time (time required
for the potential to reach the maximum) depends on the kind of
cephalosporins, the flow rate, and the activity of the bacteria-
collagen membrane. Figure 4 shows the response curves of the
microbial electrode. The maximum potential was attained in about
10 min at a flow rate of 2 ml/min. When 0.5 mM of phosphate
buffer was pumped through the reactor, the potential of the pH
electrode decreased gradually and returned to its initial level
within 20 min at 37°C.

Calibration. Figure 5 shows the relationship between the
concentration of various cephalosporins and the potential
difference (difference between the initial and the maximum
values). A linear relationship was obtained between the
logarithm of the cephalosporin concentration and the potential
difference. As shown in Fig. 5, phenylacetyl-7ADCA, cephalo-
ridine, cephalothin and cephalosporin C were determined by the
microbial electrode. The slope of the calibration curve depended
on the kind of cephalosporin employed. This may be due to the
difference in reactivity of cephalosporinase with various kinds
of cephalosporins.
    The effect of ionic strength on the potential of the glass
electrode was examined using sodium chloride. The potential of
the glass electrode was constant below 0.2 M of sodium chloride.
    The reproducibility was determined with a sample solution
containing 125 γ/ml of phenylacetyl-7ADCA and was found to be
20 $\pm$ 2 mV (10 % of the relative standard deviation) in 10
different experiments.

Reusability of the Electrode. Reusability of the microbial
electrode was examined with the sample solution containing 125 γ
/ml of phenylacetyl-7ADCA. The results obtained are shown in
Figure 6. The cephalosporin determination was performed several
times a day. However, no decrease of the potential difference
was observed for a week. This system was applied to the determination of cephalosporins in a fermentation broth. Cephalosporin C in the cephalospolium acremonium broth was determined by
the conventional method using high pressure liquid chromatography
(HPLC) (13) and the electrochemical method. The results obtained
are shown in Table 1. The relative error of the determination by
the microbial electrode was within 10 %. Therefore, this system
can be used for the continuous determination of cephalosporin in
a fermentation broth.
    To study the stability of the immobilized cephalosporinase,
they were stored in physiological saline at 5°C. Cephalosporinase in immobilized whole cells was active for a month.
    In conclusion, the determination of cephalosporin is possible

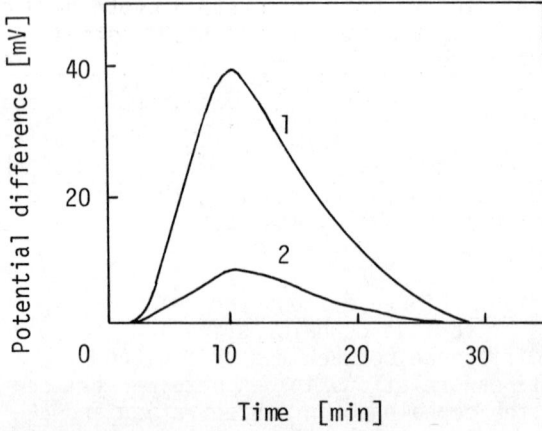

Figure 4.   *Response curve of the cephalosporin sensor. The concentration of Phenylacetyl-7ADCA was: 1. 250 γ/mL; 2. 62.5 γ/mL.*

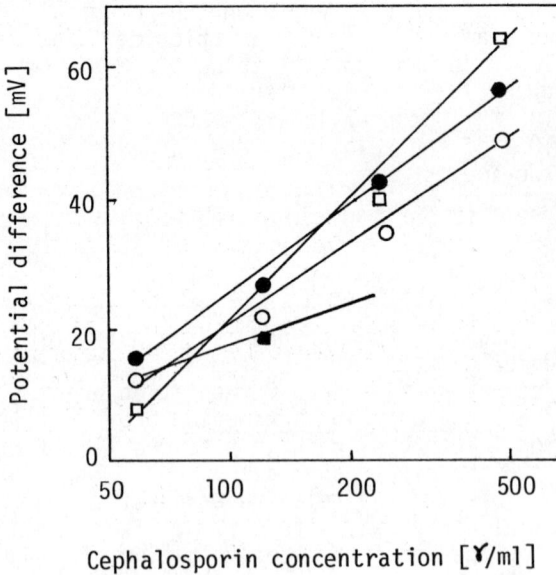

Figure 5.   *Calibration curves of the cephalosporin sensor. (□) Phenylacetyl-7ADCA, (■) cephalosporin C, (○) CET, (●) CER.*

Table I.   Determination of cephalosporin C
in the fermentation broth by microbial sensor and HPLC* method.

Method	broth [μl]	potential difference [mV]	analytical value [γ/ml]
Microbial sensor	100	7.5 ± 0.7	3500
HPLC*	5		3800

* High pressure liquid chromatography

within 10 min by using the microbial electrode. Furthermore,
cephalosporin produced in a fermentation broth can be measured
continuously by the electrode.

Microbial Electrode Sensor for Glucose

In recent years, many analytical methods involving enzymatic
reactions have been developed by using immobilized enzyme and
electrochemical devices. Enzyme electrodes using enzyme-collagen
membranes have been developed by the authors (14,15, 16,17,18).
Techniques for immobilizing living microorganisms in collagen
membrane also have been developed and these immobilized microorganisms have been applied to microbial electrodes (4,5,6).
These microbial electrodes have a potential application in the
fermentation industry. The control of fermentation process is
important for effective production of useful materials.

Molasses is used as a main carbon source in many
fermentations. Most of the carbohydrates such as glucose,
fructose and sucrose in molasses can not be determined by spectro-
photometric methods because test samples are not optically clear.
Many reports on enzyme electrodes for these carbohydrates have
been published (15,19,20,21). However, enzyme electrodes are not
suitable for determination of these carbohydrates because enzymes
are unstable and inhibitors of enzymes sometimes exist in a
fermentation broth.

Whole cells of Pseudomonas fluorescens which utilized
mainly glucose were chosen and immobilized in collagen membrane.
The microbial electrode consisting of a bacteria-collagen
membrane and an oxygen electrode for glucose is described here.

Apparatus and Procedures. The organism used in this study
was Pseudomonas fluorecences. The bacteria were cultured under
aerobic conditions at 30°C for 20 h in 80 ml of medium (pH 7.0)

containing 1 % (w/v) glucose, 1 % peptone, 1 % beef extract, and
tap water. The cells were centrifuged at 5°C and 6000 g and were
washed twice with water.

The suspension for preparation of the bacteria-collagen
membrane contained 1.8 g collagen fibrils and 0.6 g wet cells.
The bacteria-collagen membrane was prepared by casting the
suspension on a Teflon plate and dried at 20°C. The bacteria-
collagen membrane was tanned with 0.1 % glutaraldehyde solution
for 1 min and dried at 4°C.

A schematic diagram of the microbial electrode system is
presented in Figure 7. The electrode consists of a double
membrane of which one layer is the bacteria-collagen membrane
(thickness 40 μ) and the other is an oxygen permeable Teflon
membrane (thickness 27 μ), together with an alkaline electrolyte,
a platinum cathode, and a lead anode. The double membrane is in
direct contact with the platinum cathode and is tightly secured
to the cell with rubber rings. The microbial electrode was
inserted into a sample solution (20 ml), saturated with dissolved
oxygen and stirred magnetically while measurements were taken.

Glucose was determined independently by the method of Borel
et al (22) and enzymatically (23).

Response Curves of the Electrode. The bacteria in collagen
membrane were living and had the respiration activity. Table II
shows the selectivity of the electrode to various carbohydrates
and amino acids. The electrode responded slightly to fructose,
galactose, mannose and saccharose. However, the electrode did
not respond to other carbohydrates and acids. Therefore, the
microbial electrode can be applied to the determination of
glucose.

Figure 8 shows the response curves of the electrode. The
current at time zero is that obtained in a sample solution
saturated with dissolved oxygen. The bacteria began to utilized
glucose in a sample solution when the electrode was inserted into
a sample solution. Then, consumption of oxygen by the bacteria
in collagen membrane began which caused a decrease in dissolved
oxygen concentration around the membrane. As a result, the
current of the electrode decreased markedly with time until a
steady-state was reached. The steady-state indicated that the
consumption of oxygen by the bacteria and the diffusion of oxygen
from the solution to the membrane were in equilibrium. The
steady-state current was attained within 10 min. at 30°C.

When the electrode was removed from the sample solution and
placed in a solution free from glucose, the output of the microbial electrode gradually increased and returned to initial level
within 15 min at 30°C. (The current means the steady-state
current hereinafter.) The response time of the microbial
electrode was longer than that of the enzyme electrode. This may
be caused by the time lag of the bacterial respiration. However,
employment of a rate assay improved the response time and the

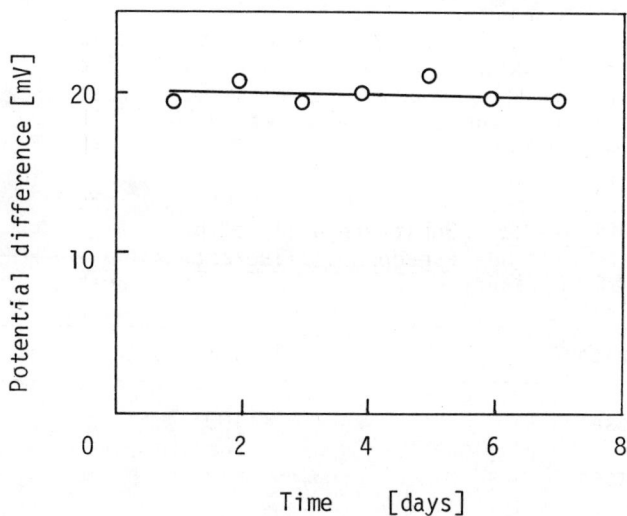

Figure 6.   Reusability of the cephalosporin sensor. Phenylacetyl-7ADCA (125 γ/mL) was employed for experiments.

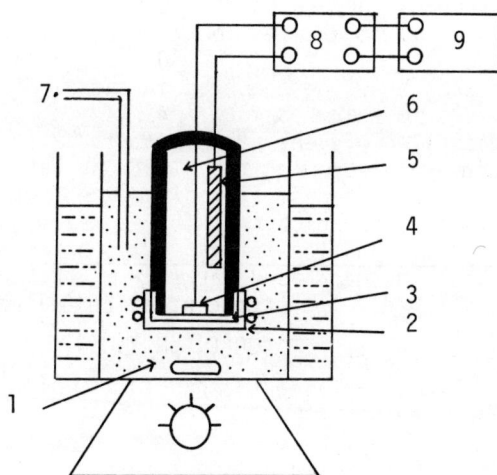

Figure 7.   Scheme of the microbial electrode sensor for glucose. 1. Sample solutions. 2. Bacteria–collagen membrane. 3. Teflon membrane. 4. Cathode (Pt). 5. Anode (Pb). 6. Electrolyte (KOH). 7. Air pump. 8. Amplifier. 9. Recorder.

Table II. Substrate utilized by
           Pseudomonas fluorescens

Substrate	Relative response
Glucose	100
Fructose	4
Galactose	9
Mannose	4
Arabinose	0
Xyrose	0
Maltose	0
Saccharose	4
Glutamic acid	0
Casaminoacids	0

Concentration 0.1 mM
30°C, pH 7.0

glucose was determined within 2 minutes.

Effect of pH and Temperature.   As the activity of bacteria
is markedly dependent on pH and temperature, these effects were
examined.

Figure 9 shows the current-pH curve for 56 μM glucose
solution at various pH values. A flat pH-current relationship
was obtained between pH 7 and 8. The current of the microbial
electrode increased below pH 7 and above pH 8. This might be
due to the inactivation of bacteria in the collagen membrane at
lower and higher pH values.

The influence of temperature on the microbial electrode was
also investigated. The amount of dissolved oxygen in water
decreases with rising temperature. Therefore, the results were
corrected to take this into account. The current output of the
microbial electrode above 40°C increased with rising temperature,
because the bacteria in collagen membrane were inactivated by
heat. As heat caused the bacteria-collagen membrane to shrink,
the exact current could not be determined above 50°C. On the
other hand, as the respiration activity of the bacteria was low
below 30°C, the current increased with decreasing temperature.

The maximum current decrease (the maximum respiration
activity of the bacteria) was observed in the neutral pH range
and in the temperature range betwen 30 and 40°C. Therefore, the
bacteria in collagen membrane are activity under these conditions.

Calibration.   The relationship between the current and
glucose concentration is shown in Figure 10. A linear relationship was obtained below a concentration of 20 mg/l. At higher
concentrations (above 100 μM), the current-concentration plots
curved toward the concentration axis. The reproducibility was
determined using the same sample (50 μM) and was found to be 32
± 2 μA (6 % of the relative standard deviation).

The assay range of the microbial electrode was from 10^{-4} to
10^{-5} M of glucose. Therefore, the sensitivity of the microbial
electrode was slightly better than that of the enzyme electrode
(19,20,21).

Determination of Glucose in Molasses.   Molasses was employed
in these experiments. It was diluted and the glucose concentration was determined by the microbial electrode and the enzymatic
method (23). Table III shows a comparison of the glucose concentrations thus determined. As shown in Table III, relative error
of the determination was within 8 %. As molasses contained
sucrose, the microbial electrode might respond slightly to sucrose
in molasses.

Reusability of the Electrode.   Reusability of the electrode
was tested as follows. The current of the microbial electrode
was determined 10 min after it had been immersed in a 50 μM

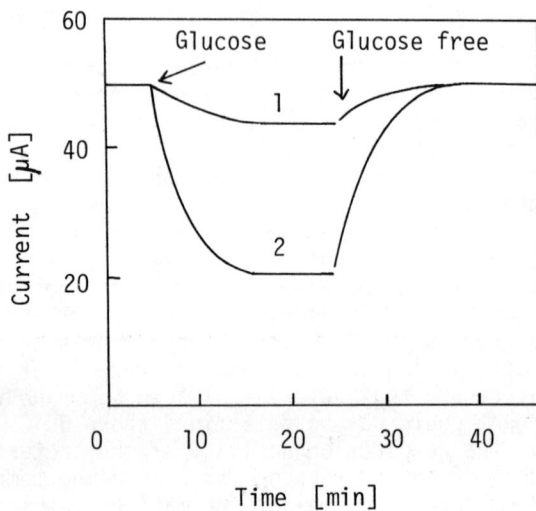

Figure 8.   *Response curves of the glucose sensor.   The concentration was: 1. 20 μM, 2. 90 μM.*

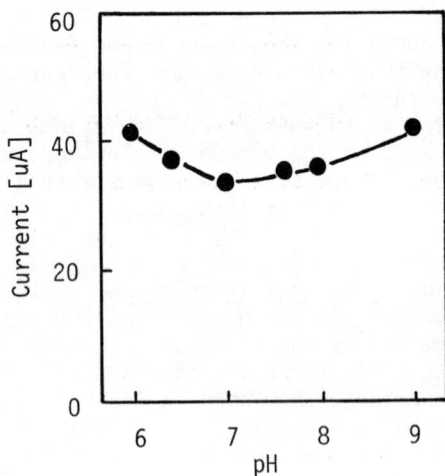

Figure 9.   *Current–pH curve of the glucose sensor*

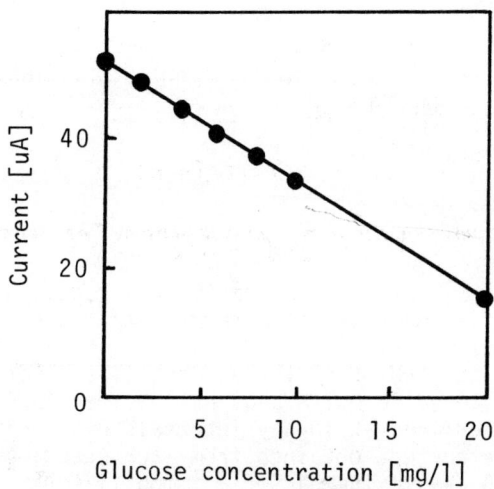

Figure 10.   Calibration curve of the glucose sensor

Table III. Determination of glucose in
            molasses by the microbial
            electrode and the enzymatic method.

Method	Glucose (%)
Microbial electrode	20.0 ± 0.8
Enzyme method*	18.5 ± 0.5

* M.E.Washka and E.W.Rice, Clin.Chem., 7,
  542 (1961)

glucose solution. The above operation was repeated three or four
times per day. No decrease in current output was observed over a
two week period.
    To study the stability of the immobilized bacteria, the
electrode was stored in 0.1 M phosphate buffer (pH 7.0) at 5°C.
Glucose was determined at 10-day intervals with this stored
electrode. The current obtained from each experiment was
constant for 30 days. The bacteria immobilized in collagen
membrane are therefore active for a month.
    In conclusion, the determination of glucose is possible
within 10 min by using the microbial electrode consisting of a
bacteria-collagen membrane and an oxygen electrode. Furthermore,
glucose in molasses can be determined by the electrode thus
rendering it possible to apply the electrode to fermentation
broths.

Other Microbial Electrode Sensors

    Many microbial electrodes have been constructed and used in
the determination of assimilable sugars, acetic acid, ethyl
alcohol, cell population, vitamin B_1, nicotinic acid, glutamic
acid and the estimation of biochemical oxygen demand (BOD). The
characteristics of these microbial electrodes are summarized in
Table IV. These microbial electrodes may be apllied to
fermentation industries in near future.

Table IV Various microbial electrodes for fermentation

Sensor	Immobilized bacteria	Device	Range (mg/l)	Response time (min)	Stability (days)
Glucose	P.fluorescens	Oxygen electrode	3-20	10	14
Assimilable sugars	B.lactofermentum	Oxygen electrode	20-200	10	20
Acetic acid	T.brassicae	Oxygen electrode	10-100	15	30
Ethyl alcohol	T.brassicae	Oxygen electrode	3-30	15	30
BOD	Activated sludge	Oxygen electrode	3-30	20	30
Cell Population	—	fuel-cell	$10^{10}-10^{11}$*	15	60
Vitamin B_1	(L.fermenti)	fuel-cell	$10^{-3}-5 \times 10^{-2}$	360	60
Cephalos- porins	C.freundii	pH electrode	60-500	10	7
Nicotinic acid	L.arabinosus	pH electrode	$5 \times 10^{-2}-5$	60	30
Glutamic acid	E.coli	CO_2 electrode	8-800	5	15

(   ) Free cells   * Numbers/l

Acknowledgements

We are grateful to Mr. K. Matsumoto, H.Seijo for their help
in this work.

Literature Cited

1. Guilbault, G.G., In Handbook of Enzymatic Methods of Analysis, Marcel Dekker Inc., New York, N.Y. (1977).
2. Chang,T.M.S., Ed., In Biomedical Applications of Immobilized Enzymes and Proteins. Vol. 2. Plenum Publizhing Co., New York, N.Y. (1977).
3. Vandamm,E.J., Chem.Ind.(1976) 1070-1072.
4. Karube,I., Matsunaga,T., and Suzuki,S., J.Solid-Phase Biochem. (1977) 2, 97-104.
5. Karube,I., Matsunaga,T., Mitsuda,S., and Suzuki,S., Biotechnol. Bioeng., (1977) 55, 243-248.
6. Karube,I., Mitsuda,S., Matsunaga,T., and Suzuki,S., J.Ferment. Technol., (1977) 55, 243-248
7. Rechnitz,G.A., Kobos,R.K., Richechel,S.J., and Gebauer,C.R., Anal.Chim.Acta, (1977) 94, 357-365
8. Kobos,R.K., and Rechnitz,G.A., Anal.Lett., (1977) 10 (10) 751-758.
9. Matsunaga,T., Karube,I., and Suzuki,S., Anal.Chim.Acta, (1978) 98, 25-30.
10. Karube,I., Suzuki,S., Kinoshita,S., and Mizuguchi,J., Ind.Eng. Chem.Prod.Res.Develop., (1971) 10, 160-163
11. Venkatasubramanian,K., and Vieth,W.R., Biotechnol.Bioeng. (1973) 15, 583-588.
12. Perret,C.J., Nature (1954), 1012-1014.
13. Miller,R.D., and Neuss,R., J.Antibiotics, (1976) 29, 902-913.
14. Aizawa,M., Karube,I., Suzuki,S., Anal.Chim.Acta, (1974) 69, 431-437.
15. Satoh,I., Karube,I., and Suzuki,S., Biotechnol.Bioeng., (1976) 18, 269-272.
16. Suzuki,S., Karube,I., and Namba,K., In Analysis and Control of Immobilized Enzyme Systems, p 151-163, North-Holland Pub. Co., Amsterdam, (1976).
17. Satoh,I., Karube,I., and Suzuki,S., J.Solid-Phase Biochem., (1977) 2, 1-7.
18. Satoh,I., Karube,I., and Suzuki,S., Biotechnol.Bioeng., (1977) 19, 1095-1099.
19. Guilbault,G.G., and Lubrano,G.O., Anal.Chim.Acta, (1975), 64, 439-455.
20. Guilbault,G.G. and Nanjo,M., Anal.Chim.Acta, (1974) 73, 367-373.
21. Tran-Mihn,G. and Broun,G., Anal.Chem., (1975) 46, 1359-1364
22. Borel,E., Hostettler,F., Deuel,H., Helv.Chim.Acta, (1952) 35, 115-126.
23. Washka,M.E., and Rice,E.W., Clin.Chim., (1961) 7, 542-551.

RECEIVED April 19, 1979.

16

Hollow Fiber Entrapped Microsomes as a Liver Assist Device in Drug Overdose Treatment

P. R. KASTL, W. H. BARICOS, and W. COHEN

Department of Biochemistry, Tulane University School of Medicine, New Orleans, LA 70112

R. P. CHAMBERS

Department of Chemical Engineering, Auburn University, Auburn, AL 36830

More than one million poisonings occur annually in the United States (1) and in 1975 over 26,000 people died from intake of medicinal and non-medicinal substances (2). Common current treatment methods include support of heart and lung function and removal of unabsorbed drug by stomach lavage. However, any drug absorbed into the bloodstream can be removed only by hemodialysis or hemoperfusion (3). These processes are expensive, require highly trained personnel, and are usually available only at large urban medical centers.

We are investigating the feasibility of using the NADPH-dependent drug hydroxylation system present in liver microsomes (μS) (fragmented endoplasmic reticulum) in an extracorporeal hollow fiber drug detoxifier (EDD) to detoxify absorbed drug in overdose patients. In the presence of NADPH and molecular oxygen microsomes will hydroxylate both endogenous as well as foreign compounds. In addition to its potential use as an EDD this system serves as a model for the use of membrane-bound, cofactor-requiring, multienzyme complexes in industrial as well as medical applications. This work also may contribute toward the development of other liver assist devices as well as provide for production of drug metabolites for subsequent therapeutic screening.

The proposed therapeutic system is shown in Figure 1. This system differs from conventional hollow fiber artificial kidney hemodialysis in several respects. First, it contains the multienzyme detoxification system in the dialysate side of the hollow fiber dialyzer, eliminating the need for large amounts of dialyzing fluid. Second, the hollow fiber device contains fibers permeable to drugs and oxygen so that both can diffuse from the blood into the μS suspension. Following transformation by the μS enzymes, the modified toxin can reenter the bloodstream, where it will be subsequently removed and excreted. An important advantage of using animal liver microsomes for detoxification in place of purified enzymes is the broad substrate specificity of the complex. Thus such a system will be applicable to drugs developed in the future as well as those in use today. We have chosen to use p-nitroanisole (pNA) and hexobarbital as model drugs in the

0-8412-0508-6/79/47-106-237$05.00/0
© 1979 American Chemical Society

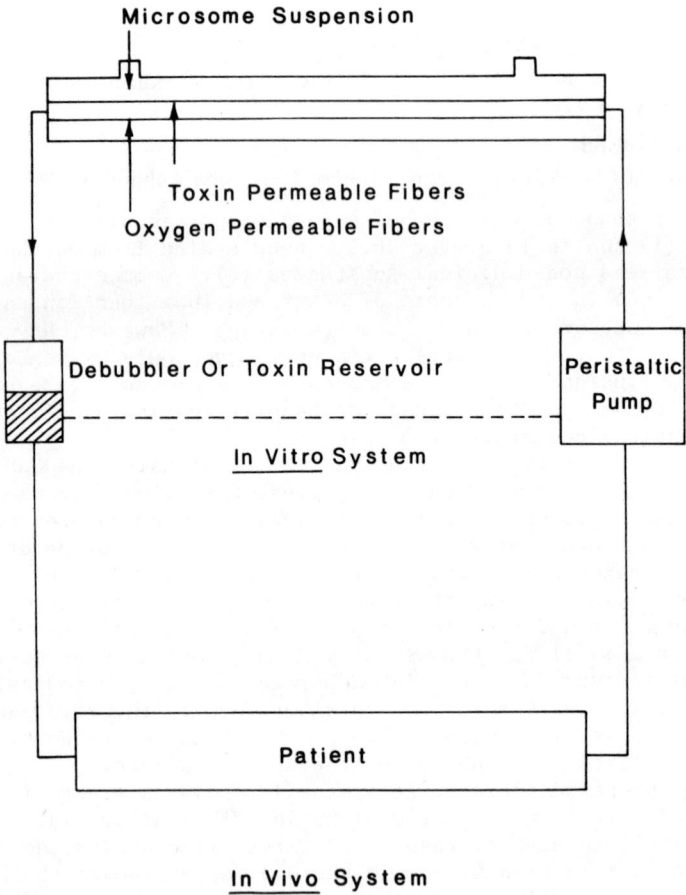

Figure 1.   Proposed extracorporeal hollow fiber drug detoxifier (EDD)

development of the EDD: pNA becuase its metabolism O-demethylation can be continuously monitored spectrophotometrically, and
hexobarbital, because it represents the most common class of
drugs encountered in drug overdose.

Methods and Materials

Phenobarbital induction of the rat liver μS drug hydroxylase enzymes, isolation, and lyophilization of μS were carried out
as previously described (4). Microsomes prepared by this procedure retained 85% of the activity of fresh nonlyophilized μS when
assayed with hexobarbital and 55% of the original activity toward
pNA. The specific activity was 9 nmoles/min/mg μS protein with
hexobarbital and 0.9 nmoles/min/mg μS protein with pNA. Microsomal pNA O-demethylase activity was assayed by a modification of
the method of Zannoni (5, 6). Hexobarbital metabolism was monitored by the method of Kupfer and Rosenfeld (7) or by the following procedure utilizing ^{3}H-hexobarbital (6). Aqueous samples
(5 μl) were extracted with ether and unmetabolized hexobarbital
was separated from its more polar metabolites by thin layer chrom-
ography (benzene/acetic acid, 9:1; silica gel G TLC plates). The
TLC plates were sprayed sequentially with diphenyl carbazone and
mercurous nitrate, and the spots corresponding to hexobarbital
and its metabolites were scraped from the plate, eluted with
methanol, and counted in a liquid scintillation counter. Protein
was determined by the method of Lowry (8). Glucose-6-phosphate
dehydrogenase (G6PDH), glucose dehydrogenase (GDH), and NADP were
purchased from Sigma Chemical Co. pNA was purchased from Aldrich
Chemical Co. and recrystallized before use. Hexobarbital sodium
was purchased from Winthrop Laboratories. ^{3}Hexobarbital was purchased from New England Nuclear. The following hollow fiber
devices were purchased from Amicon Corp. or Bio-Rad Laboratories:
Amicon Vitafiber, 50,000 nominal molecular weight cutoff, 60 cm^2
fiber surface area, acrylic copolymer fibers; Bio-Fiber 80/5 Cell
Culture Tube, 30,000 nominal molecular weight cutoff, 50 cm^2
fiber surface area, half silicone/polycarbonate fibers, half
cellulose acetate fibers.

Results and Discussion

Storage Stability of Microsomes. Prolonged storage stability under relatively mild conditions (e.g., room temperature)
is essential for any practical EDD. Microsomal suspensions retain
activity for only a few weeks at -20°C and for longer periods in
ultra-low temperature freezers at -85°C (9), the latter not being
commonly available outside of laboratories. In an attempt to
circumvent this problem we investigated the effect of lyophilization on the activity and stability of μS. Lyophilized μS prepared
from non-induced rat liver and stored in evacuated containers at
both 23°C and -20°C show no decrease in O-demethylase activity at
either temperature after 14 months of storage (Table I). The

TABLE 1

STABILITY OF LYOPHILIZED LIVER MICROSOMES AT 23°C AND -20°C

TIME AFTER ISOLATION	O-DEMETHYLASE ACTIVITY[1,2]		HEXOBARBITAL HYDROXYLASE ACTIVITY[1,3]
	23°C	-20°C	-20°C
1 day	0.26	0.39	-
1 month	0.36	0.30	-
4 months	0.36	0.30	-
6 months	0.30	0.39	-
9 months	-	-	3.83
11 months	-	0.31	3.00
14 months	0.34	0.33	-

1. nmoles/min/mg microsomal protein
2. O-demethylase activity was continuously monitored at 420 nm in a Cary 16K spectrophotometer in an assay solution containing 0.42 mg/ml microsomal protein, 1.2 mM p-nitroanisole, 2.4 mM nicotinamide, and 0.13 mM NADPH in 0.1 M sodium phosphate buffer, pH 7.8, at 25°C.
3. Hexobarbital hydroxylase activity was monitored by the method of Kupfer and Rosenfeld in an assay solution containing 0.59 mg/ml microsomal protein, 0.77 mM hexobarbital, 1.2 x 16[6] dpm of tritium-labelled hexobarbital, 2.4 mM nicotinamide, 44 mM sodium pyrophosphate, 6 mM magnesium chloride, 6 mM glucose-6-phosphate, 4 units/ml of G6PDH, and 0.13 mM NADP in 0.1 M sodium phosphate buffer pH 7.8 at 25°C.

hexobarbital hydroxylase activity of the μS preparation stored
at −20°C was determined after 9 and 11 months of storage, and
was found to be comparable to that of freshly prepared μS. Clearly storage stability of μS will not limit the practicality of a
drug detoxifier.

Functional Stability of Microsomes.   The coenzyme NADPH is
required by μS mixed-function oxidases.   Unfortunately the μS
contain other enzymes that will oxidase NADPH or destroy it and
adversely affect drug detoxification.   The most important among
these are NAD(P)ase, NAD(P)H pyrophosphatase, and NAD(P)H oxidase.   We added the competitive inhibitors sodium pyrophosphate
and nicotinamide to reaction mixtures to block the effects of
NAD(P)H pyrophosphatase and NAD(P)ase, respectively (10, 11).
An NADPH regenerating system (glucose-6-phosphate dehydrogenase/
glucose-6-phosphate) was introduced to reverse the effect of
NAD(P)H oxidase and to maintain the coenzyme at close to 100%
NADPH during the detoxification process.   The data presented in
Figure 2 demonstrate the effectiveness of optimal concentrations
of these reagents with respect to the rates and duration of microsomal O-demethylation of pNA.   The best results were obtained
using both inhibitors combined with the NADPH regeneration system.
This reaction (C in Figure 2) proceeded at a linear rate for
40 min, at which point the oxygen of the system was depleted.
In a separate experiment when oxygen was supplied by slow stirring under a stream of air, O-demethylation continued for 12 hours
(not shown).   These data demonstrate that the effects of NAD(P)H
destroying microsomal enzymes can be minimized permitting good
functional stability of the mixed-function oxidase system.

A Biocompatible NADPH Regeneration System.   Glucose-6-phos-
phate dehydrogenase/glucose-6-phosphate (G6PDH/G6P) has a common
enzyme/substrate couple used for NADPH regeneration in μS studies.
However, because magnesium ions are required for G6PDH activity,
it would have to be supplied to the EDD at a physiologically
toxic concentration.   As a practical alternative, we examined
the enzyme glucose dehydrogenase (GDH) which does not require a
metal cofactor for activity.   GDH catalyzes the NAD(P)-dependent
oxidation of glucose, producing gluconic acid and NADPH. Important
additional advantages of this enzyme are that glucose can be
supplied from the patient's blood, and the product, gluconic
acid, is relatively nontoxic.   We compared GDH/glucose to G6PDH/
G6P in a model detoxification reaction employing pNA and found
that GDH/glucose functions well as an NADPH regeneration system.
In fact, it is slightly more effective at low enzyme levels than
G6PDH/G6P (Figure 3) and thus should be satisfactory for an EDD.

Extracorporeal Hollow Fiber Drug Detoxifier.   After a func-
tionally stable μS detoxification system had been developed, in
vitro model EDD experiments were performed as schematically out-

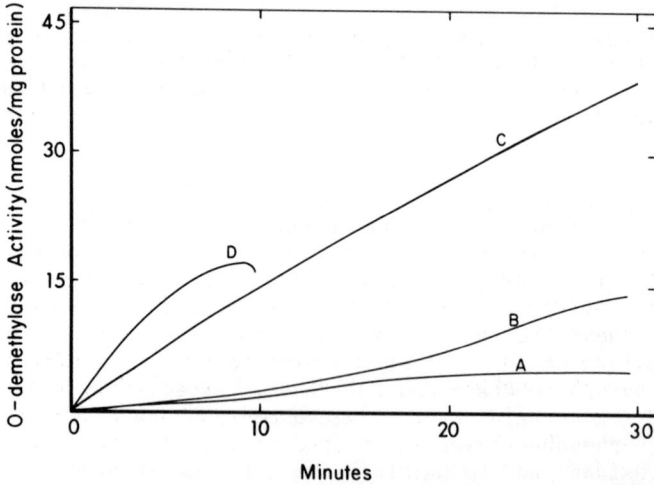

Figure 2.   Microsomal p-nitroanisole O-demethylation.   (C), Complete system containing 1.2mM pNA, 0.13mM NADP, 44mM pyrophosphate, 2.4mM nicotinamide, 6mM MgCl₂, 6mM G6P, 4 units/mL G6PDH, and 0.73 mg/mL microsomal protein in a final volume of 2.5 mL of 0.1M sodium phosphate buffer, pH 7.4; (A), Pyrophosphate and nicotinamide deleted; (B), Pyrophosphate deleted; (D), NADP, MgCl₂, and G6PDH replaced with 0.13 mM NADPH.

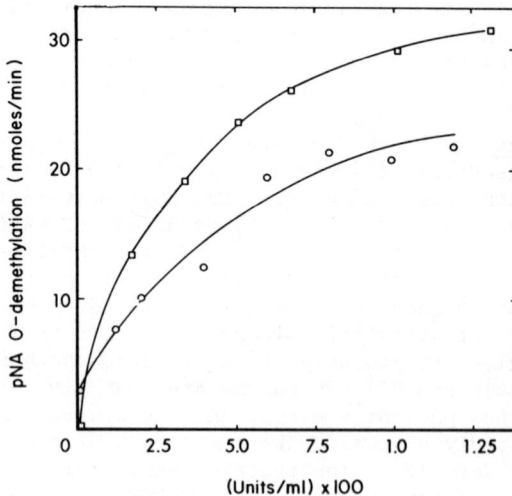

Figure 3.   Comparison of G6PDH/GDH and GDH/glucose as NADPH regeneration systems in microsomal pNA O-demethylation (○), G6PDH/G6P system as in Figure 2 (C), except 22mM pyrophosphate; (□), GDH/glucose system containing 280mM glucose, 1.2mM pNA, 0.13mM NADP, 2.4mM nicotinamide, 22mM pyrophosphate and 0.73 mg/mL microsomal protein in a final volume of 2.5 mL of 0.1M sodium phosphate buffer, pH 7.4.

lined in Figure 1.

A major problem associated with a cofactor-requiring enzyme in a hollow fiber system is retention of the relatively small cofactor within the dialysate side of the hollow fiber dialyzer (12). Increasing the molecular weight of the cofactor by convalent attachment to soluble or insoluble supports is expensive and usually results in decreased enzyme activity due to steric hindrance (13, 14). In contrast it has been demonstrated experimentally that effective cofactor reuse can be achieved without modification of the cofactor (15, 16, 17). Preliminary calculations based on a 10,000–15,000 cm^2 surface area hollow fiber reactor indicate that a slow infusion of as little as 26 mg/hr of NADP directly into the μS suspension would maintain a satisfactory NADPH concentration, thus assuring maximal rates of drug detoxification. However, in these preliminary studies adequate concentrations of NADP were included in the circulating drug solution.

It can be calculated that at ambient temperature and O_2 pressure aqueous solutions contain less than 1 μmole O_2 per ml. Even for a large volume EDD (e.g., 500 ml), the dissolved O_2 cannot support the detoxification of the amount of drug commonly encountered in overdose cases. Therefore, additional O_2 will have to be supplied. In actual use of an EDD, O_2 for detoxification would be obtained directly from the patient's blood.

In the first experiments the effects of oxygen and μS concentration on steady-state rates of hexobarbital detoxification were studied using a Bio-Fiber 80/5 cell culture tube. Figure 4 shows that oxygen bubbled through the circulating solution increased the steady-state rate of hexobarbital metabolism at all but the lowest μS concentrations. The data also show that the rate of detoxification is not directly proportional to μS concentration. Experiments measuring the rates of oxygen and hexobarbital diffusion indicate that, with 8 mg μS protein or greater, oxygen should be rate limiting, even when the solution is saturated. Diffusion of hexobarbital should not be rate limiting. The non-linearity of rate as a function of μS concentration is not as easily explained since similar results were observed with μS suspensions in test tube experiments. Two contributing factors at high μS concentrations may be aggregation of μS and inadequate protection of NADPH. Further studies will be required to resolve this question.

The rate of hexobarbital detoxification using a Bio-Fiber 80/5 cell culture tube was then compared to the rate of detoxification using an Amicon Vitafiber and to that of a μS suspension in a test tube. The steady-state rates of all three were the same (Figure 5). It is important to note that in these systems containing approximately 2.8 mg μS protein the steady-state rates were constant for the 3 hour duration of the experiments. These experiments also indicate that at this low concentration of μS no diffusional limitations exist for either hexobarbital

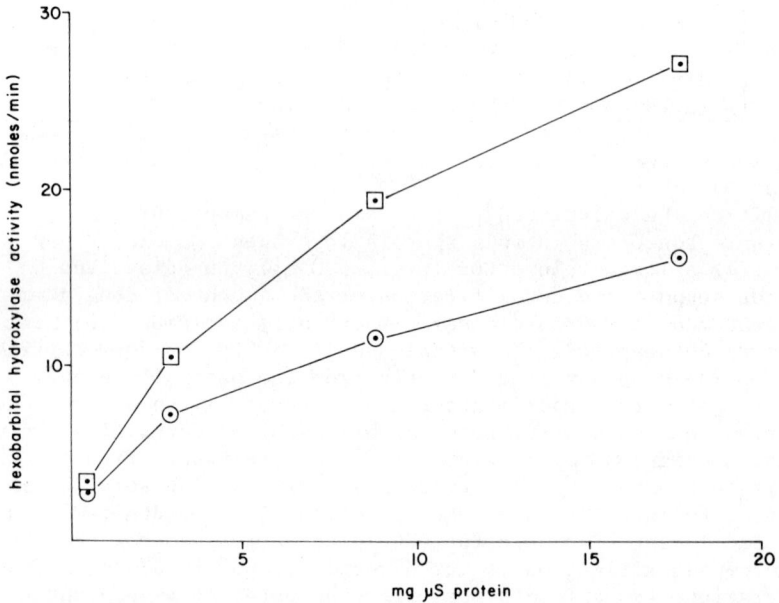

Figure 4. *Hexobarbital hydroxylase activity as a function of microsome and oxygen concentration in a Bio-Fiber 80/5 EDD. 0.5 mL of microsomes was placed on the dialysate side of the hollow fiber and 30 mL of 0.77mM hexobarbital (1.2 × 16[6] DPM ^{3}H-hexobarbital) circulated through the lumen of the fibers at 12.0 mL/min. The circulating solution also contained 0.13mM NADP, 22mM pyrophosphate, 2.4mM nicotinamide, 6mM NgCl$_2$, 6mM G6P, 2 units/mL of G6PDH, and 0.1M sodium phosphate buffer, pH 7.4. (□), Oxygen bubbled through the circulating solution; (○), ambient oxygen.*

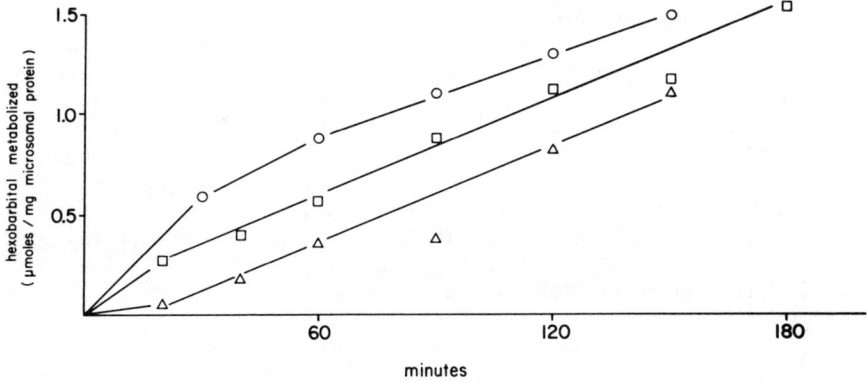

Figure 5.   Hexobarbital hydroxylase activity in a Bio-Fiber 80/5 EDD and a Vitafiber EDD compared to a simple microsome suspension as a control. The conditions were the same as in Figure 4. Oxygen was bubbled through the circulating solutions of the EDDs and the control solution was saturated with oxygen prior to the initiation of the reaction by addition of microsomes. 2.8 mg microsomal protein was used in each system. (○), Microsome suspension; (□), Vitafiber EDD; (△), Bio-Fiber 80/5 EDD.

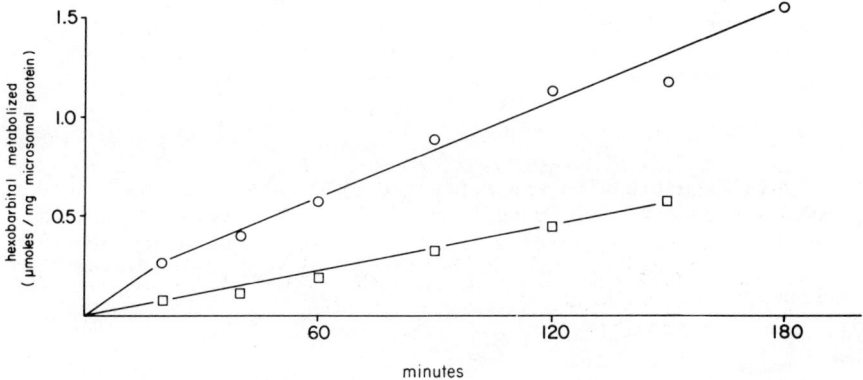

Figure 6.   Hexobarbital hydroxylase activity in a Vitafiber EDD in the presence and absence of plasma. (○), As in Figure 5; (□), plasma equilibrated with ambient oxygen replaced sodium phosphate buffer.

or oxygen.

We repeated the Vitafiber perfusion experiment using a circulating solution of hexobarbital in human plasma, equilibrated with ambient oxygen (Figure 6). The rate of hexobarbital metabolism was about 50% of that observed with a buffered oxygenated solution. This decreased rate was probably caused by mass transfer limitations secondary to binding of hexobarbital to plasma proteins and by clogging of fiber pores by the plasma proteins.

Based on the data in Figure 6 we have calculated a hexobarbital clearance rate of 0.8 ml/hr in plasma. Simple extrapolation to a clinical size hollow fiber (e.g., 10,000 cm^2 surface area) containing 17 mg μS protein/0.5 ml μS suspension (total of 4 grams of μS protein yields a theoretical clearance of 8.4 ml/min. This figure approaches the hemodialysis clearance of 15 ml/min reported for secobarbital (18), but is significantly slower than the 50 ml/min achieved using charcoal hemoperfusion (19). We feel that a μS-based hollow fiber EDD can equal and probably surpass this hemodialysis clearance through the use of more permeable membranes and improved reactor design (e.g., increased surface area, increased temperatures, more active μS, etc.). While it is doubtful that such a system can match the clearances of charcoal or resin absorption devices, the μS-based hollow fiber EDD does possess distinct advantages over these and other EDD systems. These advantages include a well established technological base (from hollow fiber renal dialysis), the elimination of immunological and throboembolic problems, and, possibly most important, a metabolic capability. We feel that this last point, a metabolic capability, is an important distinction between our μS-based EDD and physical absorption devices and extends our potential from simple detoxification to the more complex area of liver support.

ACKNOWLEDGEMENTS

This investigation was supported by grants from the Edward G. Schlieder Educational Foundation, USPHS Postdoctoral Fellowship GM00968 to PRK and Avondale Shipyards - Louisiana Heart Association. The expert technical assistance of Mrs. P. Gonzales is acknowledged.

LITERATURE CITED

1. Arena, J. W. Consultant(1977) 17 52
2. Winchester, J. R., Gelfand, M. C. Trans. Am. Soc. Artif. Intern Organs (1977) 23   762
3. Winchester, J. F., Gelfand, M. C. and Tilstone, W. J. Drug Metab. Rev. (1978) 8 69
4. Cohen, W., Baricos, W. H., Kastl, P. R. and Chambers, R. P. in "Biomedical Applications of Immobilized Enzymes and Proteins" (T.M.S. Chang, ed.). p. 319, Plenum, New York, 1977
5. Zannoni, V. G. in "Fundamentals of Drug Metabolism and Drug Disposition" (B. N. LaDu, H. G. Mandel, and E. L. Way, eds.). p. 566, William and Wilkins Co., Baltimore, 1971
6. Kupfer, D. and Rosenfeld,J J. Drug Metab. Dispos. (1973) 1 760
7. Kastl, P. R. (1978) Ph.D. Dissertation, Tulane University
8. Lowry, O. H., Rosenbrough, N. J., Farr, A. L. and Randall, R. J. J. Biol. Chem. (1951) 193 265
9. Borton, P., Carson, R. and Reed, D. T. Biochem. Pharm. (1974) 23 2332
10. Gillett, J. R., Prog. Drug Res. (1963), 6 11
11. Gillette, J. R., Grieb, W. and Sesame, H. Fed. Proc. (1963) 22 366
12. Baricos, W. H., Chambers, R. P. and Cohen, W. Anal. Letters (1976) 9 257
13. Wykes, J. R., Dunnill, P., Lilly, M. D., Biochim. Biophys. Acta (1972), 286 260
14. Lowe, C. R. and Mosbach, K. Eur. J. Biochem. (1974) 49 511
15. Baricos, W. H., Chambers, R. P. and Cohen, W. Enzyme Technol. Digest (1975) 4 39
16. Chambers, R. P., Walle, E. M., Baricos, W. H. and Cohen, W. in "Enzyme Engineering, Vol. 3" (E. K. Pye and H. H. Weetall, eds.) p. 363 Plenum, New York, 1978
17. Fink, D. J. and Rodwell, V. W. Biotechnol. Bioeng. (1975) 17 1029
18. Yaroslavskii, A. A., Firosov, N. N. and Koldaev, A. A. Byol. Eksp. Biol. Med. (1971) 71 384
19. Gelfand, M. C., Winchester, J. F., Knepshield, J. H. Hanson, K. N., Cohan, S. L., Strauch, B. S., Geoly, K. L. Kennedy, A. C., and Schreiner, G. E., Trans. Am. Soc. Artif. Intern. Organs (1977) 23 599

RECEIVED February 15, 1979.

INDEX

INDEX

251